AF553391

PROSPECTS OF NANOTECHNOLOGY

ENCYCLOPAEDIA OF NANOTECHNOLOGY-II

PROSPECTS OF NANOTECHNOLOGY

By

Dr. M.P. Arora

Dept. of Zoology
M.M.H. Post Graduate College
Ghaziabad
(U.P.)

DISCOVERY PUBLISHING HOUSE PVT. LTD.
NEW DELHI-110 002

First Published-2008

ISBN 978-81-8356-278-2

Published by

DISCOVERY PUBLISHING HOUSE PVT. LTD.
4831/24, Ansari Road, Prahlad Street,
Darya Ganj, New Delhi-110002 (India)
Phone: 23279245 • Fax: 91-11-23253475
E-mail: dphbooks@rediffmail.com
dphtemp@indiatimes.com

Printed at:

Sachin Printers, Delhi

Preface

Everyone in the modern world knows what technology is. But what is nanotechnology? Taken from the Greek, *nano* means "one billionth part of" a whole. In modern parlance, it means very, very small. In today's world of smaller cell phones and portable computers, miniaturization is the current technology. Nanotechnology is the next step after miniaturization. In tomorrow's world, nanotechnology will be the new common technology. It will affect everyone of the planet and many change civilization as we know it.

Nanotechnology is the molecular engineering. It will soon create effective machines as small as DNA. It describes the ideas and techniques that are creating or new domain of science and technology. Nanotechnology holds the promise of revolutionizing field ranging from medical science to optical communications, but requires a substantial investment in time and money to make the promise become a readily. The results could be nothing less than a new industrial revolution.

In an undertaking as new as nanotechnology, many different approaches are studied in order to ascertain the best way to fashion working devices. At present, techniques can be divided into one of two general categories. One technique, termed the top-down method, physically manipulates molecules into nanostructures. The other technique, termed the bottom-up method, coaxes atoms and molecules to assemble themselves into nanostructures.

The present title **"Prospects of Nanotechnology"** has been designed for undergraduate and post-graduate students as well as those involved in basic research in biological, biochemical and biophysical sciences.

In the preparation of this book large number of books and research papers have been consulted. So no authenticity is claimed.

The author expresses his gratitude to Mr. Wasan and staff of M/s Discovery Publishing House for their whole hearted co-operation in the publication of this book.

The author tried hard to be accurate and upto date in statement and realises the impossibility of completely avoiding errors therefore, the author will greatly appreciate having his attention called to any questionable statement.

Author

Contents

1. Introduction 1—98

Current Medical Practice, Evolution of Scientific Medicine, Volitional Normative Model of Disease, Treatment Methodology, Evolution of Bedside Practice, Changing View of the Human Body, Nanomedical Perspective, Nanomedicine and Molecular Nanotechnology, Nanomedicine: History of the Idea, Biotechnology and Molecular Nanotechnology, Naturophilia.

2. Nanoethics 99—110

Terminology and Basic Idea, Practical Development, Theoretical Development, State of the Field at Present, Utopian Dreams and Apocalyptic Nightmares, Utopian Dreams, Apocalyptic Nightmares, Toward a More Balanced Ethical View, Specific Field of Nanotechnology, Objectives of Specific Field Nanotechnology, Objectives Ethics, Development of the Field, Ethical Problems, Surmountable Ethical Problems, Conclusion.

3. Digital Technology 111—124

Great Revolution, Nanocomputers, Mechanical Nanocomputer, Storage, Reversibility, Quantum Computers.

4. Selective Applications 125—135

Nanomachinery, Scale, Shape, Processing Power, Communication, Energy, Construction, Material Properties, Bulk, Solidity, Shape, Optical Properties, Force.

5. Atomic Control 136—149

Thin-film Technology, Epitaxial Growth, Atomic Layer Epitaxy, Future Potential of Improved Multilayer Structures, Summary.

6. **Policing Science** 150—163
Lords of Creation, Catastrophic Risks, What is Different Today, Conclusion.

7. **Art of Building Small** 164—174
Electronic Industry, Lithography, Dip-pen Lithography.

8. **Long Life in an Open World** 175—188
Cell Repair Machines, Hit and Trial, Natural Efforts, Healing and Protecting the Earth, Long Life and Population Pressure, Effects of Anticipation, Progress in Life Extension.

9. **Nanotech Hobbies** 189—196
Model Railroads, Philately, Numismatics, Copying Issues, Gardening, Home Crafts, Blood Sports, Thrill Rides, Amusement Park Attractions.

10. **Human Prospect** 197—214
A World of Your Own, Elbow Room, Balance of Powers, Morality and Centralization, New Frontiers, Uploading, To Thine Own Self be True.

11. **Economics** 215—226
Robinson Crusoe, Drawbacks, Utopia, LTD, Getting Three.

12. **Space** 227—238
Getting There, Longer-term Options.

13. **Transportation** 239—250
Aircars, Airplane, Freight, Beam Me Up, Scotty.

14. **Companion: A Personal Computer** 251—264
Purpose and Scope, Design Characteristics, Appearance, Technical Details, Frames, Audiovisuals, Computing, Communications, Power, Interface, Library, Databases, Knowledge Bases, Additional Features, Alternatives, Marketing and Costs, Social Impact, Media, Politics, Everyday Life, Handicap Assistance, Barriers and Breakthroughs, Technical, Social, 2020 Vision.

15. **Nanobot Construction** 265—268

16. **Miniaturization** 269—275
Interconnection.

Index 275—283

1

INTRODUCTION

The history of disease is vastly older than that of humankind itself. Indeed, disease and parasitism have been inseparable companions to life since the dawn of life on Earth. Fossilized bacteria similar to those responsible for many infections that afflict people today have been found in geological formations that are 500 million years old. Fossil shells dating from an era almost equally remote show clear evidence of disturbance by injury and parasites. Examination of the skeletons of long-extinct dinosaurs and other great reptiles show that these creatures suffered from fractures, bone tumours, arthritis, osteomyelitis, dental caries and other diseases that still plague us in the 20th century. While available fossil evidence is largely limited to changes observable in bones and teeth, it is probably safe to assume that disease processes were equally prevalent in the soft organs and tissues that have not been geologically preserved, and that the general pattern of disease has not changed in its essentials during the hundreds of millions of years that animal life has existed on this planet.

Since its first appearance on the prehistoric stage millions of years ago, the human body has also been constantly subject to assault and injury, invasion by parasites, extremes of heat and cold, and infections. Early man probably suffered from a number of diseases due to nutritional factors and body chemistry disorders. For example, the poor condition of the teeth of an 18-year old Australopithecus who lived 1.75 million years ago, found at Olduvai Gorge by Louis Leakey in 1959, suggests that the hominid had a disease that lasted for many months—most likely gastro-enteritis due to malnutrition—with three major attacks of the disease at the ages of two, four, and four and a half. Diseases that may date back more than 25 million years to the ape ancestors of modern apes and man are amoebic dysentery, malaria, pinworm infections, syphilis, yaws, and yellow fever. Diseases which may have appeared and evolved with man include leprosy and typhoid; certain modern diseases such as cholera, measles, mumps, smallpox, whooping cough, and the common cold require large concentrated populations to support them, thus probably could not have existed in the prehistoric era.

As for cancer, Java man, first discovered by Dutch anatomist Eugene Dubois in 1891 and considered to be half a million years old, had a morbid growth on his femur. It is likely that bone cancer and other forms of cancer have existed from the earliest times. The remains of *Neanderthal*, a competing species to *Homo sapiens* that roamed through Europe, Africa, and the Near East during the last glacial period ~75,000 years ago, show clear evidence of arthritis, tooth loss, and suppurative bone disease. (The high rate of broken bones and early death suggests that *Neanderthals* engaged in more close-quarter combat with large animals than did modern humans, who had figured out safer strategies.) Human bones unearthed from the New Stone Age period reveal that Neolithic man suffered from arthritis, congenital dislocations and fractures, sinusitis, tuberculosis of the spine, and tumours.

In considering the question of how our earliest ancestors dealt with these conditions, we are on somewhat uncertain ground, since little direct evidence has been preserved. Anthropologists point out that in pre-*Neanderthal* hunter-gatherer tribes, a sick or lame person is a serious handicap to a group on the move. In the event of major illness or mortal wounds, sufferers may either leave the group, be abandoned, or, as with lepers in medieval Europe, may be ritually expelled, becoming "culturally dead" before they are biologically dead. Early hunter-gatherer hominid bands were more likely to abandon their seriously sick than to succor them, although *Cro-Magnons* and *Neanderthals* evidently were the first to care for their wounded and disabled, and to bury their dead.

But moderate wounds, bruises, fractures, or foreign bodies such as arrowheads or thorns are tangible things that demand attention. Thus the art of surgery must first have originated as a response to immediate crises. The first and most obvious course of action of a wounded man would be to protect the site of injury from the influence of external forces or agents. For this there was, and remains to this day, only one means—the application of a dressing. Many observations were made and many substances tried, and in time a body of experience was accumulated and passed on orally to others for use in similar emergencies, eventually creating a considerable sum of inherited empirical knowledge. It is believed that the art of dressing wounds long constituted the whole of medicine—the use of internal remedies or herbs, and use of the knife or fire, came much later.

Present-day primitive and folk medicines provide additional clues to early medical practice. For example, many early peoples developed effective methods to control bleeding—the use of cobwebs is an ancient folk remedy, as is the application of tourniquets, packing with absorbent materials, the laying on of snow or, at the other temperature extreme, the application of cautery by hot knife or spear. The *Masai* and *Akamba* tribes treated sword wounds by slapping on a poultice of cow dung and dust. The suturing of wounds is practiced by some primitive peoples and may have been known to prehistoric man. Bone needles furnished with an eye have been found in paleolithic deposits in France and England, and some Indian tribes suture with threads of sinew or bone needles (the needles are left in and the thread is twisted around them). One of the strangest suturing techniques, observed among primitive tribal cultures in such widely separated places as India, East Africa and Brazil, is the sealing of wounds, especially abdominal wounds, using termites or ants. The edges of the wound are drawn closely together and the insect is allowed to bite through them both, firmly securing the flesh on two sides. Once attached, the insect's body is severed, allowing only the jaws to remain in place, holding the wound shut.

While able to deal with wounds, early man did not admit the existence of disease from "natural causes". Internal diseases were generally ascribed to malevolent influences exercised by a supernatural entity or a human enemy. As centuries passed, many internal diseases were eventually recognized and simple treatments empirically established—for example, Celsus (ca. 30 AD), a Roman medical writer, described ligature (tying off blood vessels), suturing the large intestine, eye operations such as couching for cataract, tonsillectomy, and bladder surgery to remove stones. But the rational basis for internal medicine awaited basic physiological knowledge such as the circulation of the blood, finally proven by William Harvey (1578-1657) in 1628, and the discovery and acceptance of the theory of infection by microorganisms in the 1800s. Throughout most of human history, life has been short indeed. Even by the 17th century, more than half of all children never lived past the age of ten, often succumbing to diseases such as cholera, diphtheria, scarlet fever, or whooping cough, or being scarred for life by smallpox. It was a common and widely accepted fact of life that people would die at all ages, although wealthy families could leave town every year during the cholera season.

One of the constant missions of human civilization has been the avoidance and elimination of animals that prey on humans. Cave-dwelling carnivorous saber-toothed tigers, having occupied the upper echelons of the food chain for 30-35 million years, finally became extinct not more than about ten thousand years ago, most likely at the hands of newly-arrived human hunters crossing the Siberian land bridge into post-Pleistocene North America. As late as medieval times, wolves ranged freely over Europe, remaining abundant in France through 1500 AD—in winter, audacious wolf packs would enter Paris and eat children, dogs, and even adults who were alone on the streets. With technological advances, extant tiger and wolf species are now largely confined to artificial habitats or isolated nature preserves and no longer pose any serious threat to human health. People in most places are not eaten by wolves; indeed today, a few venturesome individuals actually keep and breed wolves with dogs, as pets.

Bacteria are among the last remaining "wild animals" on Earth that threaten man. As our instrumentalities continue to progress from the macroscale to the microscale, and finally to the molecular or nanoscale, all of the remaining natural "wild things" that endanger human life and health—whether viruses, bacteria, protozoa, metazoan parasites, or even our own pathological native cells—will be confined and tamed, reconstructed or eliminated. Using smaller tools, we hunt smaller prey. As with sabertooths in the post-Neolithic era, and with wolves in post-medieval times, people of the 21st century will no longer fear or need suffer predation by wild microbes or tumour cells run amok.

Humanity is poised at the brink of completion of one of its greatest and most noble enterprises. Early in the 21st century, our growing abilities to swiftly repair most traumatic physical injuries, eliminate pathogens, and alleviate suffering using molecular tools will begin to coalesce in a new medical paradigm called *nanomedicine*. Nanomedicine may be broadly defined as the comprehensive monitoring, control, construction, repair, defense, and improvement of all human biological systems, working from the molecular level, using engineered nanodevices and nanostructures.

Current Medical Practice

In order to fully appreciate the changes that nanomedicine will inevitably bring, it is useful first to review the history and development of current medical practice. As Winston Churchill once remarked: "The further backward you look, the further forward you can see." The author unapologetically favours what medical anthropologists would regard as the "Western" healing tradition. The great medical theorist Otto E. Guttentag agreed:"Contemporary Western medicine involves a type of healing that potentially and actually exceeds all other approaches in maintaining that optimal status of selfhood we call being healthy and in eliminating that reduced status of selfhood we call being sick."

Evolution of Scientific Medicine

A study of the history of "scientific" or Western medical practice suggests a continuity in certain aspects of the medical paradigm even since ancient times—as for instance the basic precepts of observation and diagnosis, followed by treatment—but reveals changes in attitudes, techniques, and instrumentalities, and a gradual evolution in the epistemology and ontology of medical logic. Note that 2010 is marked as a possible date for the first applications of *nanomedicine*. These will most likely be *ex vivo* applications only *in vivo* nanomedical treatments may come much later.

Another important reason to study the history of medicine is to gain a deeper appreciation of the long, hard struggle to improve human health, a struggle that is expected finally culminate in victory in the 21st century. If the ~10 billion people that have ever lived survived an average of 40 years and spent 5% of their lives in sickness or physical misery from disease, then ~200 trillion man-hours of suffering have been paid to achieve this remarkable result, a not inconsiderable price. There is no pretense to completeness here. For example, Chinese, Indian and Islamic contributions are omitted, not because they are unimportant, but because they did not significantly alter the evolutionary pathway of the Western medical paradigm. A great deal of physiology, pathology, neurology, systematics, and many other important medical-related disciplines are also neglected in the interest of brevity.

Prehistoric medicine

The elementary nature of prehistorical medical practice has already been mentioned. There is direct evidence that *Stone Age* man used a natural fungus as a treatment for intestinal parasites. Besides basic wound-tending, the practice of circumcision is an age-old form of simple surgery with a rational, if sometimes controversial, hygienic basis.

An even more dramatic surgical operation, for which there is considerable prehistoric fossil evidence, is trephining the skull to remove a round piece of bone from it. This can be done using flint instruments, either by gradually scratching through the skin and bones of the skull, getting gradually deeper and deeper, or by drilling a series of small holes in a circle in the skull, then cutting the small bridges between to remove a disk of bone. Skulls have been found with multiple holes. The additional fact that some hole edges show callus formation (evidence of healing) indicates survival of the patient. Trephined skulls have been found in Western Europe, including England, North Africa, Asia, the East Indies, New Zealand, and the Americas from Alaska in the north down through the continent to Peru in the south. The practice was probably regarded as therapeutic, either to remove a depressed fracture, to try to cure mental illness, or to relieve severe headache or epilepsy, presumably by letting out the demon possessing the patient.

Ancient mesopotamian medicine

The first written records, which came from ancient Babylon and Egypt, contain the earliest references to medical care,obviously codifying earlier practices no record of which has survived. The medicine of the Sumerians of Mesopotamia from ca. 3000 BC was primarily religious. The Mesopotamian peoples saw the hands of the gods in everything. Disease was caused by spirit invasion, sorcery, malice, or the breaking of taboos, and sickness was both judgement and punishment. An Assyrian text circa 650 BC describes epileptic symptoms within a demonological framework: "If at the time of the possession, his mind is awake, the demon can be driven out; if at the time of his possession his mind is not so aware, the demon cannot be driven out." Headaches, neck pain, intestinal ailments and impotence were read as omens. The appropriate remedy was to identify the demons responsible and expel them by spells or incantations.

But medicine also had an empirical component, with some sicknesses being ascribed to cold, dust and dryness, putrefaction, malnutrition, venereal infection and other natural causes. The Babylonians drew on an extensive pragmatic materia medica—some 120 mineral drugs and twice that number of vegetable items are listed in surviving tablets. Alongside various fats, oils, honey, wax, and milk, were many active ingredients that included mustard, oleander and hellebore (a plant in the buttercup family that is a violent gastro-intestinal poison, hence acts as a powerful purgative, though it is lethal in high doses). Colocynth, senna and castor oil were used as laxatives, while wound dressings were compounded with dried wine dregs, salt, oil, beer, juniper, mud or fat, blended with alkali and herbs. With the discovery of distillation, the Mesopotamians made essence of cedar and other volatile oils. Turpentine, asafetida, henbane, myrrh, mint, poppy, fig, and mandrake are also mentioned. Dog dung and other fecal ingredients were used to drive off demons.

Surgical conditions such as wounds, fractures and abscesses were also treated by Mesopotamian surgeons. Practitioners were priests, and after 2000 BC, they were ruled by the strict laws included in the Code of Hammurabi. The Code laid down rewards for success and severe punishment for failure, and contained laws relating to medical practice which show that medicine and surgery were highly organized professions. Fees were regulated on a sliding scale of rewards based on the patient's rank, and severe penalties were laid down for failure:

"Concerning the wounds resulting from operations it is written: if a physician shall produce on anyone a severe wound with a bronze operating knife and cure him, or if he shall open an abscess with the operating knife and preserve the eye of the patient, he usually shall receive 10 shekels of silver [more than a craftsman's annual pay]; if it is a slave, his master shall usually pay 2 shekels of silver to the physician."

"If a physician shall make a severe wound with an operating knife and kill him, or shall open an abscess with an operating knife and destroy the eye, his hands shall be cut off."

"If a physician shall make a severe wound with a bronze operating knife on the slave of a free man and kill him, he shall replace the slave with another slave. If he shall open an abscess with a bronze operating knife and destroy the eye, he shall pay the half of the value of the slave."

The Hammurabic Code also mentions the Gallabu, or barber-surgeons, whose province was minor surgery, including dentistry and the branding of slaves. If Herodotus (ca. 485-425 BC) may be believed, Babylonian medicine must have declined in the 5th century BC. Herodotus states that there were no physicians, but that the people brought their sick into the marketplace in order that passers-by might make suggestions or offer cures.

Ancient Egyptian medicine

The civilization of ancient Egypt dates from around 3000 BC. As in Mesopotamia, ancient Egyptian medicine was religious-empirical. An examination of the preserved bodies of members of the Royal family has provided much information on the diseases of ancient Egypt, including congenital deformities such as clubfoot, dental caries, gallstones, bladder and kidney stones, rheumatoidarthritis, mastoiditis, numerous eye diseases, and bone fractures, some of which show treatment by quite sophisticated splinting. Other evidence suggests that the average ancient Egyptian was extremely diseased. For example, a weaver who died in the 11th century BC, aged 14-18 years, evidently suffered from *schistosomiasis*, tapeworm likely associated with malnutrition, *anthracosis* of the lungs presumably due to environmental pollution from cooking and heating, *pulmonary silicosis*, and possibly malaria and fleas.

The Ebers Papyrus (ca. 1550 BC), deriving from Thebes, is the principal medical document and may be the oldest surviving medical book. Over 20 meters long, it deals with scores of diseases and proposes remedies involving spells and incantations, but also includes many rational treatments. The Ebers Papyrus covers 15 diseases of the abdomen, 29 of the eyes, and 18 of the skin, and, perhaps unsurprisingly to the modern consumer, lists no fewer than 21 cough treatments. About 700 drugs and 800 formulations are mentioned, mainly involving herbs but also including mineral and animal remedies. For example, to cure night blindness the patient should eat fried ox liver—possibly a tried-and-tested procedure, since liver is rich in Vitamin A, lack of which causes the illness. Eye disorders were also common, for example:

"To drive away inflammation of the eyes, grind the stems of the juniper of Byblos, steep them in water, apply to the eyes of the sick person and he will be quickly cured. To cure granulations of the eye, prepare a remedy of cyllyrium, verdigris, onions, blue vitriol, powdered wood, and mix and apply to the eyes."

For stomach ailments, a decoction of cumin, goose-fat and milk was recommended, but other remedies sound more exotic, including a drink prepared from black ass testicles. A mixture of vulva and penis extracts and black lizard was supposed to cure baldness. Also good for hair growth was a compound of hippopotamus, lion, crocodile, goose, snake and ibex fat—merely assembling these ingredients might promise a hair-raising experience! Egyptian medicine credited many vegetables and fruits with healing properties, and also tree products such as sycamore bark and resins such as myrrh, frankincense and manna. As in Mesopotamia, plant extracts—notably senna, colocynth and castor oil—were employed as purgatives, and oil of camomile to improve digestion. Recipes included ox spleen, pig's brain, stag's horn, honey-sweetened tortoise gall, and the blood and fats of various animals.

Antimony, copper salts, alum, and other minerals were recommended as astringents or disinfectants. Containing ingredients from leeks to lapis lazuli—including garlic, onion, tamarisk, cereals, spices, condiments, resins, gums, dates, hellebore, opium and cannabis—compound drugs were administered in the form of pills, ointments, poultices, fumigations, inhalations, gargles and suppositories; they might even be blown into the urethra through a tube. Among the most interesting of the healers whose names have been recorded for posterity are Peseshet, head female physician or overseer, proof of the existence, as in Mesopotamia, of female healers; and Iri, Keeper of the Royal Rectum, presumably the pharaoh's enema expert.

The Edwin Smith Papyrus (ca. 1600 BC), discovered by its American namesake at Luxor in 1862, may be the world's earliest surviving surgical text, and contains material probably derived from even more ancient times. The "*book of wounds*" comprises 48 case reports, which commence with the top of the head and proceed systematically downward —nose, face, ears, neck and chest, mysteriously stopping in mid-sentence at the spine, presumably when the scribe was interrupted at his work. The only surgical conditions treated were wounds, fractures, abscesses, and circumcisions. In the Papyrus, the method of presentation is first to set out the title or the chief symptom, followed by the further symptoms, and then the examination, diagnosis, prognosis, and treatment. The advice given is entirely rational, as will be seen by the following directions for the treatment of a fractured humerus:

"Instructions concerning a Break in his Upper Arm. If thou examinest a man having a break in his upper arm, and thou findest his upper arm hanging down, separated from its fellow, thou shouldst say concerning him: One having a break in his upper arm. An ailment which I will treat."

"Thou shouldst place him prostrate on his back, with something folded between his two shoulder blades; thou shouldst spread out with his two shoulders in order to stretch apart his upper arm until that break falls into its place. Thou shouldst make for him two splints of linen, and thou shouldst apply for him one of them both on the inside of his arm, and the other of them both on the underside of his arm. Thou shouldst bind it with ymrw (an unidentified mineral substance), and treat it afterward with honey every day until he recovers."

An interesting point in the case histories is that after the diagnosis the writer gives a decision about his further course of action. The verdict may take one of three forms: (i) an ailment which I will treat;

(ii) an ailment with which I will contend; and (iii) an ailment not to be treated. Like modern military triage, the "hopeless" patient is left to his inevitable fate. This guarded attitude on the part of the medical man was widespread in antiquity. While present-day doctors generally do everything possible to alleviate symptoms to the very end, even when the patient has no chance of recovery, the view in ancient times was that hopeless cases were not to be touched. This attitude was entirely practical. A doctor in attendance at the courts of ancient Egypt might expect rich rewards if his patient recovered, but if a patient died under his care, the unfortunate physician ran a grave risk of impalement.

Ancient Greek medicine

By 1000 BC the communities later collectively known as the Greeks were emerging around the Aegean Sea. How much medical knowledge they took from Egypt remains controversial, but the contrasts between the two are striking. Little is known of Greek medicine before the appearance of written texts in the 5th century BC. Archaic Greece had its folk healers, including priest healers employing divination and herbs. From early times (the first Olympic games were recorded in 776 BC), the love of athletics produced instructors in exercise, bathing, massage, gymnastics and diet. The Homeric epics (ca. 600 BC) offer glimpses of early Greek medicine. Scholars count 147 cases of battle wounds in the Iliad, including 106 spear thrusts, 17 sword slashes, 12 arrow shots, and 12 sling shots. Among the arrow wound survivors was King Menelaus of Sparta, whose physician extracted the arrow, sucked out the blood and applied a salve. As with other medical interventions in Homer, this shows no Egyptian influence, supporting the idea that, even if Greek practice owed much to Egypt, it rapidly went its own way.

Various Greek gods and heroes were identified with health and disease, the chief being *Asclepius*, who even had the power to raise the dead. A heroic warrior and blameless physician, Asclepius was the son of Apollo, sired upon a mortal mother, who was taught herbal remedies by Chiron and then generously used them to heal humans. Incensed at being cheated of death, Hades, the ruler of the underworld, appealed to the supreme god, Zeus, who obligingly dispatched Asclepius with a thunderbolt, though he was later elevated to godhood. A different version appears in Homer, who portrays Asclepius as a tribal chief and a skilled wound healer whose sons became physicians and were called Asclepiads, and from whom all Asclepian practitioners descended. As the tutelary god of medicine, Asclepius is usually portrayed with a beard, staff and snake—the origin of the caduceus symbol of the modern physician, with its two snakes intertwined, double-helix like, on a winged staff. The god was often shown accompanied by his daughters, Hygeia (health or hygiene) and Panacea (cure-all).

For all that, *Hippocratic medicine*, the foundation of Greek written medicine, explicitly grounds the art upon a quite different basis—a healing system independent of the supernatural and built upon natural philosophy. The beginnings of true medical science in the West were established when the reliance on superstition that underpinned tribal medicine was replaced by civilized and rational curiosity about the cause of illness. The growth of civilized thought allowed for argument on medical cause and cure, with great doctrinal multiplicity. The separation of medicine from religion reveals another distinctive feature of Greek healing: its openness, a quality characteristic of Greek intellectual activity in general, owing to political diversity. There was no imperial Hammurabic Code and, unlike *Egypt*, no state medical bureaucracy, nor were there examinations or professional qualifications. Those calling themselves doctors (iatroi) had to compete with bone-setters, exorcists, root-cutters, incantatory priests, gymnasts, and showmen, exposed to the quips of playwrights and the criticism of philosophers. Medicine was open to all.

Empedocles (fl. 450 BC) may have been the first to advance some of the key physiological doctrines in Greek medicine, including innate heat as the source of living processes such as digestion, the cooling function of breathing, and the notion that the liver makes the blood that nourishes the tissues. His contemporary, *Alcmaeon of Croton* (fl. 470 BC), believed that the brain, not the heart, was the chief organ of sensation. Alcmaeon's examination of the eyeball led him to discern the optic nerve leading into the skull, a genuine observational basis. He gave similar explanations for the sensations of hearing and smelling, because the ear and nostrils suggested passages leading to the brain. Most of such knowledge depended heavily on wound observation and animal dissection, for in the classical period the dignity of the human body forbade dissection.

All we know for sure of *Hippocrates* (ca. 460-377 BC), who "taught all that were prepared to pay," is that he was born on the island of Cos and lived a long and virtuous life. The sixty or so works comprising the *Hippocratic Corpus* (ca. 440-340 BC) derive from a variety of hands, and, as with the books of the Bible, they became jumbled up, fragmented, then pasted together again in antiquity. What is now called the *Corpus* was gathered around 250 BC in the Library at Alexandria, with further texts added later still. Some volumes are philosophical, others are teaching texts or case notes. What unites them all is the conviction that health and disease are capable of explanation by reasoning about nature, independently of supernatural interference. Man is governed by the same physical laws as the cosmos, hence medicine must be an understanding, empirical and rational, of the workings of the body in its natural environment. Anticipating modern medicine, appeal to reason, rather than to rules or to supernatural forces, gives Hippocratic medicine its distinctiveness. It was also patient- rather than disease-centered; the Hippocratics specialized in medicine by the bedside, prizing trust-based clinical relations:

"Make frequent visits; be especially careful in your examinations, counteracting the things wherein you have been deceived at the changes. Thus you will know the case more easily, and at the same time you will also be more at your ease. For instability is characteristic of the humours, and so they may also be easily altered by nature and by chance."

"...Keep a watch also on the faults of the patients, which often make them lie about the taking of things prescribed. For through not taking disagreeable drinks, purgative or other, they sometimes die. What they have done never results in a confession, but the blame is thrown upon the physician."

The Hippocratics promulgated the idea of "*vis medicatrix naturae*," or the power of nature to cure itself, and thus the belief that there was a natural tendency for things to get better on their own. This tendency could be aided by providing a beneficial environment for the patient and by improving physical function with a regimen of suitable diet, lifestyle, and exercise—the diatetica. In extreme cases, further aids to recovery could be sought. Stubbornly offending "*humors*" could be removed with the help of venesection (phlebotomy or bloodletting) and purgatives, sudorifics applied to induce sweating, and diuretics to increase urination. But the Hippocratic physician was extremely reluctant to administer drugs of any kind, because the physician's goal was to aid nature in healing the body.

Hippocratics scorned heroic interventions and left risky procedures to others. Their Oath explicitly forbade cutting, even for stones, and other texts reserved surgery for those used to handling war wounds. Surgery was regarded as an inferior trade, the work of the hand rather than the head, a fact reflected in its name: "surgery" derives from the Latin "*chirurgia*," which comes from the Greek "cheiros" (hand) and "ergon" (work); surgery was handiwork. Hippocratic surgical texts were thus conservative in outlook, encouraging a tradition in which doctors sought to treat complaints first through management, occasionally through drugs, and finally, if need be, by surgical intervention.

The art of diagnosis involved creating a profile of the patient's lifestyle, work and dietary habits, partly by asking questions and partly by the use of trained senses:

"When you examine the patient, inquire into all particulars; first how the head is...then examine if the hypochondrium [abdomen beneath lower ribs] and sides be free of pain, for...if there be pain in the side, and along with the pain either cough, tormina [painful intestinal colic] or bellyache, the bowels should be opened with clysters [enema].... The Physician should ascertain whether the patient be apt to faint when he is raised up, and whether his breathing is free..."

Hippocratics prided themselves on their clinical acuity, being quick to pick up telltale symptoms, as with the facies hippocratica, the facial look of those dying from long-continued illness or cholera: "a protrusive nose, hollow eyes, sunken temples, cold ears that are drawn in with the lobes turned outward, the forehead's skin rough and tense like parchment, and the whole face greenish or black or blue-grey or leaden." Experience was condensed into aphorisms, as for instance: "When sleep puts an end to delirium, it is a good sign."

The technique most prized among Hippocratics was the art of prognosis—a secular version of the priestly and oracular prognostications of earlier medicine, and bearing some analogy to the 20th century weatherman, who can give a bright or gloomy forecast but is powerless to change it. Noted one Hippocratic text:

"It appears to me a most excellent thing for the physician to cultivate Prognosis; for by foreseeing and foretelling, in the presence of the sick, the present, the past, and the future, and explaining the omissions which patients have been guilty of, he will be the more readily believed to be acquainted with the circumstances of the sick; so that men will have confidence to entrust themselves to such a physician....Thus a man will be the more esteemed to be a good physician...from having long anticipated everything; and by seeing and announcing beforehand those who will live and those who will die."

The ultimate significance of Hippocratic medicine was twofold. First, it carved out a lofty role for the selfless physician which would serve as a lasting model for professional identity and conduct. Second, it taught that an understanding of sickness required an understanding of nature.

Ancient Alexandrian medicine

Soon after the death of *Aristotle* (d. 322 BC) and his most famous pupil, *Alexander* the Great (d. 323 BC), a great medical school was founded in Egypt at the court of King Ptolemy, at his capital, *Alexandria*, at the mouth of the Nile. The King's main cultural creations were the Alexandrian Library and the Museum (Sanctuary of the Muses), which installed Greek learning in a new Egyptian environment—Archimedes, Euclid, and the astronomer Ptolemy were soon to teach there. The Library became a wonder of the scholarly world, eventually containing, it was said, 700,000 manuscripts, and other facilities including an observatory, zoological gardens, lecture halls and rooms for research.

The two earliest teachers at the Alexandrian medical school were also its greatest—*Herophilus* of Chalcedon (ca. 330-260 BC) and his contemporary, *Erasistratus* of Chios (ca. 330-255 BC). Their writings having been lost, we know about them only through later physicians.

Herophilus was the first to dissect cadavers in public. He was a student of *Praxagoras* of Cos (fl. 340 BC), who had improved *Aristotelian* anatomy by distinguishing arteries from veins, but who saw the arteries as air tubes, similar to the trachea and bronchi, a common error because arteries are devoid of blood in corpses. Herophilus observed that the coats of arteries were much thicker than those of the veins, thus he speculated that the arteries were filled not with air but with blood. *Herophilus* wrote at least eleven treatises, discovering and naming the prostate and the duodenum (from the Greek for twelve fingers, the length of gut he found). He also wrote on the pulse as a diagnostic guide and on therapeutics, ophthalmology, dietetics, and midwifery. He recognized the brain as the central organ of the nervous system and the seat of intelligence, extending the knowledge of the parts of the brain, certain of which still bear titles translated from those given by him. He was the first to grasp the nature of the nerves, which he distinguished as motor and sensory, though he did not separate them clearly from tendons.

Erasistratus surmised that every organ is formed of a threefold system of "vessels" —veins, arteries, and nerves, dividing indefinitely. These, plaited together, were postulated to make up the tissues. In the brain, *Erasistratus* observed convolutions, noting that they were more elaborate in man than in animals and associating this with higher intelligence. He distinguished between cerebrum and cerebellum, and is often regarded as an early mechanist because of his model of bodily processes—for instance, digestion involved the stomach grinding food. He was opposed to intrusive remedies such as venesection, and his favourite therapeutic measures were regulated exercise, diet, and the vapour bath—very much in the Hippocratic tradition.

Ancient Roman medicine

Roman tradition held that one was better off without doctors. According to Cato (234-149 BC), citizens had no need of professional physicians because Romans were hale and hearty, unlike the effete Greeks. Apparently Romans enjoyed bad-mouthing Greek physicians. Thus the Romans despised medicine as a profession but this did not prevent them from making use of Greek physicians or even of the services of their own slaves. *Galen of Pergamum* (130-200 AD) tells us that in his time, large cities such as *Rome* and *Alexandria* swarmed with specialists who also travelled about from place to place. *Martial* (40-104 AD) mentions some of them in an epigram: "Cascellius extracts and repairs bad teeth; you, Hyginus, cauterize ingrowing eyelashes; Fannius cures a relaxed uvula without cutting; Eros removes brand marks from slaves; Hermes is a very Podalirius for ruptures." Under the Empire, military medicine was highly organized—every cohort had its surgeon, and surgeons of a higher grade were attached to the legions as consultants. Army surgeons ranked as noncombatants and enjoyed many privileges.

In the Roman empire, the earliest scientific teacher was a Greek, Asclepiades of Bithynia (124-40 BC). *Asclepiades* ridiculed the Hippocratic expectant attitude as a mere "meditation on death," and urged active measures that the cure might be "seemly, swift and sure." Though outside the mainstream, his medical practice is interesting as a modification of the atomic or corpuscular theory, according to which disease results from an irregular or inharmonious motion of the corpuscles of the body. His pupils were numerous, constituting the Methodical school, but his available therapeutic tools were few—he trusted mainly to changes of diet, friction, bathing, exercise, and occasionally emetics, bleeding, and wine. He was also the first to use music in the treatment of the insane.

Back in the mainstream, *Galen of Pergamum* (130-200 AD) provided the final medical synthesis of antiquity and the effective medical standard for the next 13 centuries. His first medical appointment was as surgeon to the Roman gladiators. He later traveled to Rome and wrote extensively on anatomy, physiology and practical medicine. His fame is due in part to his prolific pen some 350 authentic titles ranging in topic from the soul to bloodletting polemics survive, about as much as all Greek medical writings together. Galen was a flamboyant character. One of his party tricks, revealing his genius for self-advertisement as well as experiment, was to sever the nerves in the neck of a pig. As these were severed, one by one, the pig continued to squeal; but when Galen cut one of the laryngeal nerves the squealing stopped, impressing the crowd.

Galen justified venesection in terms of his elaborate pulse lore. Written in the early 170s, his sixteen books on the pulse were divided into four treatises, each four books long. In one of these treatises, he explains how to take the pulse and to interpret it, raising key questions, for example: How was it possible to tell whether a pulse was full, rapid, or rhythmical? Such questions he resolved partly from experience and partly by reference to earlier authorities. Galen developed a characteristic physiological scheme that remained in vogue until the 17th century. It supposes three types of so-called spirits associated with three types of the activity of living things. These were the natural spirits formed in the liver and distributed by the veins, the vital spirits formed in the heart and distributed by the arteries, and the animal spirits formed in the brain and distributed by the nerves. Galen's system was an admirable if factually flawed working hypothesis, based on much experimental evidence, and he presented his work as "perfecting" the legacy of Hippocrates.

As in Greece, medicine remained personal in Rome. No medical degrees were conferred or qualifications required. In the absence of colleges and universities, the private face-to-face nature of medical instruction encouraged fluidity and diversity, Students attached themselves to an individual teacher, sitting at his feet and accompanying him on his rounds. Many different sorts of medical care were available, and self-help was universal. Celsus' *On Medicine* was written for a non-professional readership as willing to wield the scalpel as the plough. Disease explanations changed little. Public authorities still ascribed famines and pestilences to the gods, and Galen was silent on contagion. The essential trilogy of classical medicine remained dietetics, exercise, and drugs, accompanied by light surgery.

Medicine in the Middle ages

The passage from the glorious days of Rome to the Middle Ages was often violent, especially in the West, with wave after wave of barbarian onslaughts from the East. These culminated in the sack of the Eternal City by Alaric's Goths in 410 AD, which effectively ended the western empire and frayed the thread of learned medicine. Thus Galen had no effective successor. Indeed, medieval medicine may be summed up as a corrupted version of Galenism. The true scientific tradition did not reappear in the West until the 16th century, after a lengthy incubation in the Islamic world. For example, in England the Venerable Bede (ca. 672-735) and his monks possessed many medical writings, and knowledge of plant remedies was extensive, but the English healer used chants and charms, predicated on the belief that certain diseases and bad luck were caused by darts shot by elves, while other ailments involved a "great worm," a term applied to snakes, insects, and dragons.

By contrast with the naturalist focus of *Hippocratic* and *Galenic medicine*, healing became more authoritarian and intertwined with religion, for the rising Church taught that there was a supernatural plan and purpose to everything, including sickness and death. *Christian* and *Jewish* healing traditions became more prominent. Disease could be cured by prayers or by invoking the names of saints, by exorcism, by amulets or number magic, or by transferring the sickness to animals, plants, or to the soil. Certain maladies such as leprosy were associated with the Almighty's punishments for sin, according to the Book of Leviticus:

"When a man shall have in the skin of his flesh a rising [a swelling], a scab, or bright spot, and it be in the skin of his flesh like the plague [the spots] of leprosy...and the plague in sight be deeper than the skin of his flesh it is a plague of leprosy: and the priest shall look on him, and pronounce him unclean."

Such polluting diseases were curable by the Lord alone, encouraging some Jewish people to reject human medicine in favour of the divine, citing the fate of *King Asa* (ca. 914-874 BC), who "sought not the Lord, but his physicians," and whose foot sores consequently worsened, and he died. On the other hand, while *Jewish* dietary rituals (e.g. kosher food) are principally expressions of religious precepts about pollution and purification, a useful practical effect is to limit exposure to foodborne diseases such as trichinosis.

Many of the distinguished surgeons of the Middle Ages were clerics, but the practice of medicine and surgery by members of the Church was not favoured by the hierarchy. Many laws were passed against the practice of medicine for worldly profit. In 1215 the Fourth

Lateran Council forbade all sub-deacons, deacons, or priests to practice that part of surgery that had to do with burning and cutting. Finally, Pope *Honorius III* (d. 1216) prohibited all persons in holy orders from practicing medicine in any form, which meant that the educated classes were prohibited from performing any type of surgery.

In Europe in the Middle Ages, practitioners unsuccessfully sought to cure wounds by treating only the weapons that caused them, using an ointment on the offending knife or sword known as a weapon salve. Physicians practiced urine-gazing (uroscopy), but proper chemical analysis of urine did not develop until the 18th century. Medieval medicine is the source of many humorous "remedies" in contemporary diatribes against modern medicine—such as being strapped into a halter and walked around a pigsty three times for mumps, drinking water from the skull of a bishop to treat rheumatism, or nuzzling a mouse to cure the common cold. But we must recognize that such treatments were not put forth as reasonable science. Rather, they serve as a reminder of the results of injecting mystical or religious elements into medicine as a substitute for good science.

The Middle Ages are also remarkable for their plagues, which became more numerous as people crowded closer and closer together in urban centers of increasing population density. *Plague* records from China during this period often show 50%-70% mortality rates, and China probably served as the original source of the greatest of the bubonic plagues that swept through Europe in the 14th century, known as the Black Death. The disease broke out in 1346 among the armies of a Mongol prince who laid siege to the trading city of Caffa in the Crimea. This compelled his withdrawal, but not before the disease had entered Caffa, whence it rapidly spread by ship throughout the Mediterranean.

The initial shock in 1346-1350 was severe. Die-offs varied widely, with some small communities experiencing total extinction and others, such as Milan, being spared entirely. The lethal effect of the plague may have been enhanced by the fact that it was propagated not solely by the bites of fleas carried on infected rodents, but also was transmitted person to person as a result of inhaling bacillus-filled droplets that had been coughed or sneezed into the air by an infected individual. Lung infections of this kind were observed to be 100% lethal in Manchuria in 1921, the only time modern medicine has directly observed airborne plague communication, so it is tempting to assume a similar mortality for pneumonic plague in 14th century Europe.

Mortality rates for sufferers from bubonic infection transmitted by flea bite varies from 30% to 90%. All told, the best estimate of European plague-induced mortality is that about one-third of the population died during the initial five-year period.

The plague returned in the 1360s, the 1370s, and thereafter, and European population declined irregularly as a result, reaching a low point in England sometime between 1440-1480. People learned to minimize infection risk via quarantine, a practice which stemmed from Biblical passages prescribing the ostracism of lepers. Plague sufferers were treated as though they were temporary lepers, with a standard 40-day quarantine for people and incoming ships at all major ports. However, since the role of fleas and rats in disease propagation remained unknown until the end of the 19th century, quarantine measures were often ineffectual. Through the 17th century, occasional plague outbreaks that carried off up to a third or a half of a city's population in a single year were considered normal. For example, Venetian statistics show that in 1575-77 and again in 1630-31, a third or more of the city's population died of plague.

Pre-modern medicine was powerless against bubonic plague. Before antibiotics reduced the disease to triviality in 1943, the average mortality rate was 60%-70% of those affected, despite all that the best hospital care could accomplish. The last recorded outbreak of plague that ran its course without benefit of penicillin and related antibiotics (which destroy the infection rapidly) occurred in Burma in 1947, with 78% lethality.

Renaissance and Pre-modern medicine

While internal medicine languished through the Middle Ages, the surgeons of the Renaissance gained wide experience during the many religious wars of the period. They had many new problems to face, including the treatment of wounds caused by firearms. The French had employed gunpowder at the siege of Puiguillaume in 1338, and cannon were used by the English at the Battle of Crecy in 1346. Gunshot wounds at that time were caused by large missiles of low velocity that caused ragged wounds and carried pieces of clothing into the tissues. These wounds were severe and very liable to become septic. The universal belief among contemporary surgeons was that gunpowder itself was venomous. To neutralize the effect of this venom, the general practice was to cauterize the wound by injecting boiling oil.

The first man to break away from this old doctrine was *Ambroise Pare* (1510-90). Pare came to Paris in 1532 as an apprentice to a barber-surgeon and then moved to the great Hotel Dieu as resident surgeon. In that immense medieval hospital, the only one in Paris at the time, he gained great experience, and in 1536 he began his career as a military surgeon. As described in his book *The Apologie and Treatise*, he recounts how, during his first campaign as a greenhorn military surgeon in Turin in 1537, he had run out of boiling oil, the established treatment for gunpowder wounds, just after French troops had captured the castle of Villaine. So in place of boiling oil, he applied:

"...a digestive of yolk of eggs, oil of roses, and turpentine. In the night I could not sleep in quiet, fearing some default in not cauterizing, that I should find those to whom I had not used the burning oil dead impoisoned; which made me rise very early to visit them, where beyond my expectation I found those to whom I had applied my digestive medicine, to feel little pain, and their wounds without inflammation or tumour, having rested reasonable well in the night; the other to whom was used the said burning oil, I found them feverish with great pain and tumour about the edges of their wounds. And then I resolved with myself never so cruelly to burn poor men wounded with gunshot."

Pare also went on to show that bleeding after amputations should be arrested, not by the terrible method of the indiscriminate use of the red-hot cautery, but by simple tying of the blood vessels. His most famous phrase, so reminiscent of the old Hippocratic school of thought, was: "I dressed the wound, and God healed him."

In 1633 appeared the earliest book on first-aid for the injured, by one *Stephen Bradwell*, although the proffered advice sounds impractical and a bit odd to modern ears. For example, the treatment for the "Biting of a Madde Dogge" is to throw the patient into water. "In doing this, if he cannot swim, after he hath swallowed a good quantity of water, take him out again. But if he be skilful in swimming, hold him under the water a little while till he have taken in some pretty quantity." This procedure may not be wholly irrational—standard 20th century first aid for dog bites includes a thorough cleansing of the wound with water.

Venesection remained a popular 17th century universal remedy. As described by the surgeon *Richard Wiseman* (1622-1676), "a gentleman of about thirty years of age coming out of *Hertfordshire* through Tottenham and riding upon the causeway near an inn, one emptying a chamber pot out of the window as he was passing by, his horse started and rushed violently between a signpost and a tree which supported part of the sign. The poor gentleman was beaten off his horse and lay stunned upon the ground." A barber-surgeon was hastily summoned but nothing much was done for the injured man until Wiseman arrived, whereupon:

"I found the gentleman lying upon the ground, the people and chirurgeon gazing upon him. I felt his pulse much oppressed, the right brow bruised and inquired whether they had bled him blood. The chirurgeon replied that he had opened a vein in his arm but it would not bleed. I replied, we must make him bleed through it by splitting his veins. Turning his head on one side, I saw the jugular vein on the bruised side turgid and opened it. He bled freely. After I had taken about twelve ounces, the blood ran down from his arm which had been opened before and would not bleed. We bled him till he came to life, and then he raved and struggled with us."

The patient's injuries were dressed and he was subjected to further bleedings, but evidently made a good recovery.

Even by the 18th century, the traditional surgeon's day-to-day business eschewed high-risk operations like amputations; rather, it was a round of minor procedures such as venesection, lancing boils, dressing skin abrasions, pulling teeth, managing whitlows, trussing ruptures, and treating skin ulcers. The fatality rates of these procedures were low, for surgeons understood their limits, and the repertoire of operations they attempted was small, because of the well-known risks of trauma, blood loss, and sepsis. Internal disorders were treated not by the knife but by medicines and management, since major internal surgery was unthinkable before anesthetics and antiseptic procedures. Improvements did occur in certain operations such as lithotomy. *William Cheselden* (1688-1752), a great British surgeon of the 18th century, perfected a technique which enabled him to remove a stone in the bladder in one minute (his record time was 54 seconds), thus reducing mortality from about 50% to under 10%. Cheselden's results were not bettered until almost the end of the 19th century.

In the 17th century, internal medical treatment was frequently overdone on those affluent enough to afford it. Critics often denounced physicians as meddlesome, capriciously practicing an often dangerous polypharmacy—a blunderbuss approach. The deathbed of *Charles II* (1630-1685) of England was a conspicuous case of such medical overkill; after the king had suffered a stroke, his doctors moved in, and *Sir Raymond Crawfurd* (1865-1938) recreated the scene:

"Sixteen ounces of blood were removed from a vein in his right arm with immediate good effect. As was the approved practice at this time, the King was allowed to remain in the chair in which the convulsions seized him. His teeth were held forcibly open to prevent him biting his tongue. The regimen was, as Roger North pithily describes it, first to get him to wake, and then to keep him from sleeping. Urgent messages had been dispatched to the King's numerous personal physicians, who quickly came flocking to his assistance; they were summoned regardless of distinctions of creed and politics, and they came. They ordered cupping-glasses to be applied to his shoulders forthwith, and deep scarification to be carried out, by which they succeeded in removing another eight ounces of blood. A strong antimonial emetic was administered, but as the King could be got to swallow only a small portion of it, they determined to render assistance doubly sure by a full dose of Sulphate of Zinc. Strong purgatives were given, and supplemented by a succession of clysters. The hair was shorn close, and pungent blistering agents were applied all over his head. And as though this were not enough, the red-hot cautery was requisitioned as well."

One of the dozen attending physicians noted with pride that "nothing was left untried"; the King graciously apologized for being "an unconscionable time a-dying."

Meanwhile, the common citizen experienced poor health exacerbated by the many new dangers attending the Industrial Revolution. In 1775, *Percivall Pott* (1714-1788) pointed out that boy chimneysweeps developed scrotal cancer, due to soot irritation. In his *Condition of the Working Classes in England* (1844), *Friedrich Engels* (1820-1895), a Manchester factory owner as well as *Karl Marx's* collaborator, described workers who were "pale, lank, narrow-chested, hollow-eyed ghosts," cooped up in houses that were mere "kennels to sleep and die in." In 1832, the Leeds physician *Charles Turner Thackrah* (1795-1833) published *The Effects of Arts, Trades, and Professions on Health and Longevity*, documenting the diseases and disabilities of various occupations. Apart from factory workers, among those most exposed to harmful substances were cornmillers, maltsters, coffee-roasters, snuff-makers, rag-pickers, papermakers and feather-dressers. Tailors were so subject to anal fistulas that they set up their own "*fistula clubs*." Thackrah's overall verdict was bleak: "Not 10% of the inhabitants of large towns enjoy full health."

The single worst malady cultivated in populous cities was tuberculosis (TB), a disease characterized by fever, night sweats, and hemoptysis (coughing up blood), called "consumption" because victims were almost literally consumed. By 1800 TB was proclaimed the most common disease, and in 1815 *Thomas Young* (1773-1829) surmised that tuberculosis brought a premature death to one in four in the general population. Autopsies conducted in the chief Paris hospitals recorded TB as the cause of death in some 40% of cases. In the continental U.S. as late as 1890, the corresponding percentage was about 13%, and tuberculosis was still the leading cause of death, though the data are partially suspect because cases of lung cancer were sometimes reported as "consumption".

One important 18th century improvement in internal medicine which decisively saved many lives was the introduction of inoculation and vaccination against smallpox. Smallpox, "the speckled monster," had become virulent throughout Europe and in bad years accounted for about 10% of all deaths; *Queen Mary of England* (1662-1694), *Louis XV of France* (1710-1774), and Queen Anne's son and sole surviving direct heir (d. 1700) died of it. Doctors had long been aware of the immunizing properties of an attack, and smallpox inoculation seems to have been known and practiced for centuries at a folk level throughout Arabia, North Africa, Persia, and India. Reports of a more elaborate Chinese method, involving the insertion of a suitable infected swab of cotton inside the patient's nostril, reached London in 1700. But it was a report from *Mary Wortley Montagu* (1689-1762), wife of the British consul in Constantinople, that Turkish women held smallpox parties at which they routinely performed inoculations with the aim to induce a mild dose so as to confer lifelong protection without pockmarking, that hastened acceptance in the rural medical community. The usual method was to transfer the infection by introducing matter from a smallpox pustule into a slight wound made in the patient's skin. Occasionally the patient developed a severe case of smallpox from such treatment, and some died. But usually the symptoms were slight—a few score of pox only—and immunity proved equivalent to that resulting from contracting the disease naturally.

Edward Jenner (1749-1823), an English country doctor who performed such inoculations, noticed that cowpox, a cattle disease occasionally contracted by humans, particularly dairy maids, also conferred immunity against smallpox. Suspecting that it might be possible to produce this immunity by arm-to-arm inoculation from the cowpox pustule, and surmising it would be safer than inoculation from smallpox pustules directly, since in humans cowpox was benign, Jenner tried the experiment, and it worked. In 1798 he published his discovery in *An Inquiry into the Causes and Effects of the Variolae Vaccinae*.

By 1799 over 5000 individuals had been vaccinated in England and abroad the practice was taken up remarkably swiftly, being made compulsory in Sweden and supported by *Napoleon*, who had his army vaccinated. For the first time in history, organized medicine began to contribute to human population growth in a statistically significant fashion.

The Napoleonic Wars spurred new attempts to treat battle-wounded soldiers in a more timely manner. Traditionally, the wounded were left on the field unattended until the end of battle, but Napoleon's chief surgeon *Dominique Jean Larrey* (1766-1842) introduced the use of rudimentary carts called ambulances volantes (little more than horsedrawn rickshaws) as the first "ambulances" to evacuate and transport wounded soldiers from the field to nearby aid stations, even while the battle raged on. In 1792, Larrey organized the first air evacuation, by hot air balloon.

The use of ambulances didn't catch on until the late 1800s; until then, anyone injured in the streets of Paris, London, New York or Boston depended on the kindness of strangers or a nearby business shop for a place to rest until a doctor could be summoned. The first "modern" ambulance appeared in the city of Cincinnati in 1865, but the first true city ambulance system was developed in association with Bellevue Hospital in New York City in 1866, receiving 1500 requests for transport in its first three years of service. These horse-drawn ambulances, usually provided by local mortuaries, carried a driver and a surgeon, who was on board mainly to pronounce a patient's death at the scene or upon arrival at the hospital, since little could be done for the seriously injured. The surgeon kept meticulous notes on the ride, recording the time of the call, transport and arrival times, and any other details that "a coroner's jury might possibly require".

The period also saw the development of many simple diagnostic tools that are taken for granted today. For example, a French physician, *Rene Theophile Hyacinthe Laennec* (1781-1826), in his *Treatise On Mediate Auscultation* (1819), described pathological lesions found in the chest at autopsy and showed how they correlated with disease detected in living patients, establishing for the first time the modern concept of clinicopathological correlation, the cornerstone of modern diagnosis. Laennec also developed an instrument that he named a "stethoscope" to assist him in his examination of patients, especially female patients, against whose chests the direct placing of a male ear was socially taboo. The original device was a straight wooden tube; by mid-century rubber tubing was introduced to create a flexible monaural stethoscope, and in 1852 an American physician, *George P. Cammann* (1804-1863), devised our familiar two-ear instrument.

The clinical thermometer is of like vintage. *Galileo* (1564-1642) invented the first thermometer in the late 16th century, but it was not applied to medicine. Early medical thermometers in the 18th century were a foot long and difficult to use at the bedside, and were reportedly carried under the arm "as one might carry a gun." The short clinical thermometer was devised by *Sir Clifford Allbutt* (1836-1925) in the 1860s, and was widely used during the American Civil War (1861-1865). The classical work in temperature diagnostics was *Carl Wunderlich's* (1815-1877) *The Temperature in Diseases*, published in 1868, which presented data on nearly 25,000 patients and analyzed temperature variations in 32 diseases, showing that temperature readings could differentiate fevers. Other devices emerged later to measure pulse and blood pressure. In 1854, *Karl Vierordt* (1818-1884) created the sphygmograph, a pulse recorder usable for routine monitoring on humans. Blood pressure was measured using the familiar inflatable band wrapped around the upper arm, called the sphygmomanometer, whose basic design was established in 1896 by *Scipione Riva-Rocci* (1863-1937). The hypodermic syringe was invented in 1853.

Biochemistry also began to play an increasing diagnostic function. In the 18th century, *Matthew Dobson* (1784) developed tests for diabetes. In 1827, *Richard Bright* (1789-1858) showed show the kidney complaint subsequently called Bright's disease could be diagnosed by a single, simple chemical test. Chemical analysis was crucial to Alfred Becquerel's (1814-1862) urinalysis studies in 1841, establishing the average amounts of water, urea, uric acid, lactic acid, albumin, and inorganic salts secreted over 24 hours, and correlating these with various disease conditions. In 1859, *Alfred Garrod* (1819-1907) devised a simple chemical test pathognomonic for gout.

Fully invasive surgery

The fully invasive surgery of the 19th and 20th centuries rests upon the triple foundation of anatomy, anesthesia, and asepsis. Each has an interesting story, described below, presaging the emergence of the rational-scientific approach to medical treatment.

Anatomy

In surgical practice, lack of accurate anatomical knowledge had long been a great obstacle. Dissection of the human body was first practiced systematically at the great medical school of Alexandria, which flourished from about 300 BC until the death of the last ruler of Ptolemaic Egypt, Cleopatra, in 30 BC. After the decline of Alexandria, dissection was carried on at a few other centers in the Middle East, but in the first two centuries of the Christian era human bodies were replaced on the dissection table by those of apes and other animals. The anatomical knowledge gained by Galen and others from the dissection of animals was an adequate guide for the simple operative procedures carried out at this time because the abdomen, the chest, and the head were rarely opened by the surgeon's knife.

Anatomical demonstrations of a kind were introduced into some Italian medical schools early in the 14th century but their main purpose was to serve as an aid in memorizing what Galen had written a thousand years before. During the late Middle Ages, and for a long time after, the procedure was for the professor to read from some second- or third-hand manuscript version of Galen while a demonstrator pointed to the part under discussion with a wand. As the text of Galen was often based on the dissection of an ape or pig, there were naturally many occasions when the anatomical structure under examination did not correspond with Galen's description.

Some of the great artists took up the scientific study of anatomy, but the true founder of modern anatomy was *Andreas Vesalius* (1514-1564), a native of Brussels who studied medicine in Paris and later taught surgery and anatomy at Padua and Bologna. Vesalius was filled with a passionate desire for anatomical study, and many stories are told of the great risks which he took in obtaining material, including, it is reported, graverobbing. On one occasion he stole the skeleton of a criminal which was hanging on a gallows outside the city wall of Louvain and the trophy proved of great value in his studies. His public dissections during his seven years in Padua drew enormous crowds of students. In 1543 he published his 355-page great work, *De Humani Corporis Fabrica Libri Septem (On the Fabric of the Human Body)*, the outstanding and precise illustrations in which were copied and plagiarized by scholars for more than 100 years, and could be used for teaching even today. For the first time in the history of medicine, doctors had at their disposal a detailed and accurate anatomical text with illustrations from the hand of a great artist. This book is the foundation stone of anatomy—indeed of all modern medicine—because without a sound knowledge of the structure of the body there can be no real understanding of the body's functions in health and disease.

A tremendous shortage of dead human bodies for dissection, to teach anatomy, remained a problem for centuries. In Edinburgh in

1827, an old man died in *William Hare's* (1792-1870) boardinghouse; assisted by his lodger William Burke (1792-1829), the two men bypassed the grave and sold the body directly to anatomists. Spurred by success, they turned to murder, luring victims and suffocating them to avoid signs of violence. Sixteen were done to death and their bodies sold, fetching 7 pounds apiece, before Burke and Hare were brought to justice in 1829. Hare turned King's evidence and Burke was hanged. The last cadaver was found in the dissecting room of a respected anatomist, *Robert Knox* (1791-1862). Despite his cries of innocence, an incensed crowd burned down his house and he fled to London, his career in ruins, and eventually *Knox* died in obscurity.

To help pass the English Anatomy Act of 1832 (which awarded to doctors the "unclaimed bodies" of paupers) and to dispel public concern about dissection, the body of the great English philosopher and jurist, *Jeremy Bentham* (1748-1832), in accordance with his directions, was dissected in the presence of his friends. The skeleton was then reconstructed, supplied with a wax head to replace the original (which had been mummified), dressed in Bentham's own clothes and set upright in a glass-fronted case. Both this effigy and the head are preserved at University College, London, to this day.

Anesthesia

Pain did not prevent surgery but made it almost unbearable, and the accompanying trauma often proved dangerous. Before the anesthetic era, which commenced in the late 1840s, surgical operations were agonizing. Of course, if the patient had a broken leg or amajor wound, there was no choice but submit to a surgeon's knife. But non-emergency elective operations would only be undergone if the condition itself was so painful or life-threatening that the victim could even consider allowing surgery. In this event, patients would choose a surgeon with the best reputation for quickness—a limb might be removed or a bladder stone evacuated in a couple of minutes. Progress in technique was often rapid. For example, in 1824, *Astley Cooper* (1768-1841) took 20 minutes to amputate a leg through the hip joint; ten years later, *James Syme* (1799-1870) was doing it in just 90 seconds.

In pre-anesthetic days, operations were rushed through at lightning speed and under conditions of appalling difficulty. The most hardened surgeons had to steel themselves to perform operations which they knew would cause agony to their patients and nerve-wracking distress to themselves. It is hard for any 20th century inhabitant of an industrialized nation to imagine what a major surgical operation must have meant to the patient in the days before anesthesia. The following is a personal account by a male patient who suffered the removal of a stone from the bladder by *Henry Cline* (1750-1827), surgeon to St. Thomas's Hospital and one of the leading operators of the day, on 30 December 1811, just three decades before the widespread adoption of anesthesia:

"My habit and constitution being good it required little preparation of body, and my mind was made up. When all parties had arrived I retired to my room for a minute, bent my knee in silent adoration and submission, and returning to the surgeons conducted them to the apartment in which the preparations had been made. The bandages &c. having been adjusted I was prepared to receive a shock of pain of extreme violence and so much had I overrated it, that the first incision did not even make me wince although I had declared that it was not my intention to restrain such impulse, convinced that such effort of restraint could only lead to additional exhaustion. At subsequent moments, therefore I did cry out under the pain, but was allowed to have gone through theoperation with great firmness."

"The forcing up of the staff prior to the introduction of the gorget gave me the first real pain, but this instantly subsided after the incision of the bladder was made, the rush of urine appeared to relieve it and soothe the wound."

"When the forceps was introduced the pain was again very considerable and every movement of the instrument in endeavoring to find the stone increased. Still, however, my mind was firm and confident, and, although anxious, I was yet alive to what was going on. After several ineffectual attempts to grasp the stone I heard the operator say in the lowest whisper, "It is a little awkward, it lies under my hand. Give me the curved forceps," upon which he withdrew the others. Here, I think, I asked if there was anything wrong—or something to that purport — and was reanimated by the reply conveyed in the kindest manner, "Be patient, Sir, it will soon be over." When the other forceps was introduced I had again to undergo the searching for the stone and heard Mr. Cline say, "I have got it." I had probably by this time conceived that the worst was over; but when the necessary force was applied to withdraw the stone the sensation was such as I cannot find words to describe. In addition to the positive pain there was something peculiar in the feel. The bladder embraced the stone as firmly as the stone was itself grasped by the forceps; it seemed as if the whole organ was about to be torn out. The duration, however, of this really trying part of the operation was short and when the words "Now, Sir, it is all over" struck my ear, the ejaculation of "Thank God! Thank God!" was uttered with a fervency and fulness of heart which can only be conceived....I never heard what was the precise duration of the operation but conceive it to have been between twelve and fifteen minutes."

And now, a woman's point of view. In 1810, Napoleon's famed military doctor Dominique *Jean Larrey* performed a radical mastectomy without anesthetic on the popular female novelist *Fanny Burney* (1752-1840). Burney later wrote a long account of the operation which, despite the excruciating agony, she believed had nevertheless saved her life:

"M. Dubois placed me upon the Mattress, & spread a cambric handkerchief upon my face. It was transparent, however, & I saw through it that the Bed stead was instantly surrounded by the 7 men and my nurse. I refused to be held; but when, bright through the cambric, I saw the glitter of polished steel — I closed my eyes..."

"Yet — when the dreadful steel was plunged into the breast — cutting through veins—arteries—flesh—nerves—I needed no injunctions not to restrain my cries. I began a scream that lasted unintermittingly during the whole time of the incision—& I almost marvel that it rings not in my Ears still! so excruciating was the agony."

"When the wound was made, & the instrument was withdrawn, the pain seemed undiminished, for the air that suddenly rushed into those delicate parts felt like a mass of minute but sharp & forked poniards [small pointed daggers], that were tearing at the edges of the wound. But when again I felt the instrument, describing a curve, cutting against the grain, if I may so say, while the flesh resisted in a manner so forcible as to oppose & tire the hand of the operator, who was forced to change from the right to the left —then, indeed, I thought I must have expired, I attempted no more to open my eyes....The instrument the second time withdrawn, I concluded the operation over—Oh no! presently the terrible cutting was renewed—& worse than ever, to separate the bottom, the foundation of the dreadful gland from the parts to which it adhered...yet again all was not over...."

When did the practice of anesthesia begin? The Herbal of Dioscorides (ca. 40-90 AD) contains specific directions for giving a decoction (boiled extraction) of mandragora "to such as shall be cut or cauterized," one of the earliest references to surgical anesthesia. Bernard de Gordon (ca. 1260-1308) tells us that the Salernitans rubbed up poppy seed and henbane and used them as a plaster to deaden the sensibility of a part to be cauterized. *Arnold of Villanova* (1235-1311) gives the following recipe:

"To produce sleep so profound that the patient may be cut and will feel nothing, as though he were dead, take of opium, mandragora bark, and henbane root equal parts, pound them together and mix with water. When you want to sew or cut a man, dip a rag in this and put it to his forehead and nostrils. He will soon sleep so deeply that you may do what you will. To wake him up, dip the rag in strong vinegar."

Some of the surgical textbooks of the Middle Ages contain references to anesthetic sponges which were prepared by soaking them in various herbs reputed to have soporific properties. The favourite herb for this purpose was the mandrake. Another simple method of producing analgesia used intermittently from early times was compression. Writing in 1564 about the various uses of the tourniquet, the French surgeon *Ambrose Pare* noted that "it much dulls the sense of the part by stupefying it." Amusingly, reliable witnesses claim that as late as the 19th century, a method of anesthesia practiced at the Imperial Court of China was to "knock the patient out by a sudden blow on the jaw."

The almost complete absence of any mention of pain-relieving drugs in medical literature of the post-medieval period is not easy to explain. It is, however, probable that the action of crude concoctions employed in early times was very uncertain, and that drugged sleep often ended in death. The active ingredients of the many herbs used in medicine had not been isolated and it would have been very difficult to regulate dosages reliably.

The story of inhalation anesthesia begins in 1799 when *Sir Humphry Davy* (1778-1829) recorded the effects produced by the inhalation of nitrous oxide. He breathed various concentrations of the gas and noted that a headache and the pain associated with the cutting of a wisdom tooth were relieved. Demonstrations of the effects of nitrous oxide were frequently given, bladders filled with "laughing gas" being passed around at lectures. Gas inhalation became a popular party game. A little book of 1839 contains adescription of the "irresistibly ridiculous" sight of a large room filled with persons each of whom was sucking from a bladder. As the gas began to take effect, "some jumped over the tables and chairs; some were bent on making speeches; some were very much inclined to fight; and one young gentleman persisted in attempting to kiss the ladies." At about the same time, "ether frolics" became equally popular.

In January 1842, *William E. Clarke* (b. 1818), a young American physician of Rochester, New York, who had acquired some knowledge of ether by attendance at ether frolics, administered the chemical on a towel to a Miss Hobbie who then had one of her teeth extracted painlessly. So far as is known this was the first use of ether for a dental or surgical operation. In March 1842, *Crawford W. Long* (1815-1878) of Danielville, Georgia, who had also witnessed ether frolics "enjoying sweet kisses from the girls," successfully removed a small tumour from the neck of a patient under the influence of ether. *Horace Wells* (1815-1848), a dentist of Hartford,Connecticut, attended a public demonstration of the effects of nitrous oxide in December 1844, and the day after he administered the gas to himself and had one of his own teeth pulled out by a colleague. Afterwards, Wells wrote: "I didn't feel it so much as the prick of a pin." His former partner, *William Thomas Green Morton* (1819-1868), also introduced ether into his dental practice in 1846. By February 1847, the Lancet and other medical journals were reporting anesthetic operations from all parts of Great Britain, and ether had been used in most European countries. In June 1847 the news reached South Africa and a leg was amputated painlessly by *W.G. Atherstone* of Grahamstown.

Sir James Young Simpson (1811-1870), professor of surgery at Edinburgh, introduced chloroform in 1847. Tradition has it that Simpson had been testing chemicals with his assistants when somebody upset a bottle of chloroform; upon bringing in dinner, Simpson's wife found them all asleep. Unlike ether, chloroform did not irritate the lungs or cause vomiting, and was powerful and easy to administer. In April 1853, *Queen Victoria* (1819-1901) took chloroform for the birth of Prince Leopold; *John Snow* (1813-1858) administered the anesthetic. Protests followed–some objections were religious (e.g., the Bible taught that women were supposed to bring forth in "travail and pain") but most were medical, putatively on grounds of safety but ringing with naturophilia: "In no case could it be justifiable to administer chloroform in perfectly ordinary labour," complained the Lancet. Incredibly, some early 19th century surgeons believed that using anesthesia for an operation would weaken a patient's character.

Of course, general anesthesia could indeed prove dangerous, and deep unconsciousness was unnecessary for less invasive procedures, so the search was on for substances that would numb a particular area for local surgery. Cocaine was isolated in 1859 and was first used in ophthalmologic procedures by Carl Koller (1857-1944). Cocaine became the first widely-used local anesthetic, synthesized in 1885 by the Merck drug company.

Germ theory and antisepsis

By the middle of the 19th century, pain had been banished from surgical operations, but one grave danger still faced every patient submitting himself to the surgeon's knife. This was the ever-present risk of sepsis (infection). Hospital diseases such as erysipelas, pyemia, septicemia and gangrene, were rife. In the 1850s the death rate after amputations varied from 25%-60% in different countries and in military practice it reached the appalling figure of 75%-90%. The first ovariotomies, which were the first abdominal operations performed on a fairly large scale, had a mortality rate of more than 30% even in the most expert hands. That all these diseases were due to some form of "contagion" had long been suspected, but the general view was that whatever agent was responsible was generated spontaneously in wounds. Alternatively, it was theorized that air itself was responsible for suppuration and many attempts were made to exclude the air from wounds by means of elaborate dressings.

Some medical men had postulated the existence of minute particles in the air which carried contagion, the so-called "germ theory." In 1546, *Girolamo Fracastoro* of Verona (1478-1553) proposed "seminaria, the seeds of disease which multiply rapidly and propagate their life," minute bodies passing unseen from the infector to the infected by contact, by clothing or utensils, and by infection at a distance through the air.

What made the germ theory of contagion so difficult to accept was that no one could see the supposed microbes. Magnifying lenses were used in ancient times, and by the beginning of the 17th century they had been combined in a tube to make the compound microscope. The first man to employ the microscope in investigating the causes of diseases was probably *Athanasius Kircher* (1601-1680), a learned Jesuit priest. In 1658 Kircher described experiments upon the nature of putrefaction, showing how maggots and other living creatures developed in decaying matter. He also claimed to have found in the blood of plague-stricken patients "countless masses of small worms, invisible to the naked eye." It is impossible that he could have seen the plague bacillus with the very low power microscopes at his disposal, but he may have seen some of the larger microorganisms and his statements about the doctrine of contagion are even more explicit than those of Fracastoro.

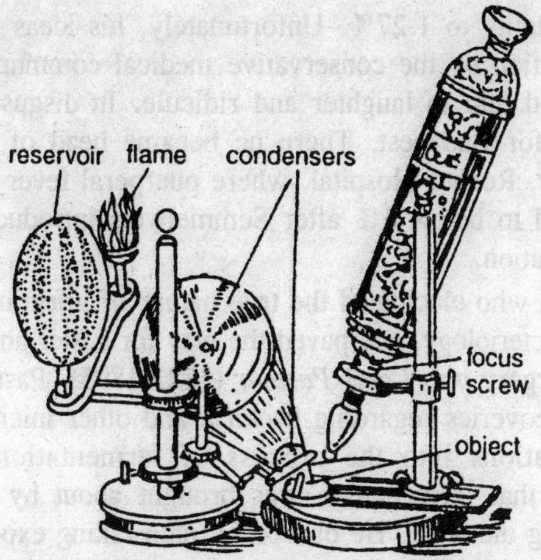

Fig. 1.1. Anthoni van Leeuwenhoek microscope

The great pioneer of modern microscopy was *Anthoni van Leeuwenhoek* (1632-1723), a Dutch linen draper, who ground his own lenses and made hundreds of microscopes. Few of his instruments provided a magnification of more than 160X but he is generally credited as making the first observations of germs, reported in his communications to the Royal Society in London. Leeuwenhoek was the first to describe spermatozoa and he gave the first complete account of the red blood corpuscles in 1674; he also found that the film from his own teeth contained "little animals, more numerous than all the people in the Netherlands."

In 1847, *Ignaz Philipp Semmelweis* (1818-1865), an assistant at the Vienna General Hospital in the maternity clinic (the world's largest at the time), was investigating the 29% postpartum mortality rate among women in Ward One, where births were handled by medical students, vs. a 3% rate in Ward Two where births were handled by midwifery pupils. Semmelweis noticed that the appearances of these deaths from puerperal fever looked the same as those observed in the body of an older colleague, forensic medicine Professor *Jakob Kolletschka* (1803-1847), who had died from adissection wound suffered in the clinic. He correctly surmised that the postpartum deaths were caused by infection from "putrid particles" carried on the hands of medical students who often shuttled back and forth between the labour wards, the obstetrical clinic, and the autopsy room where dissections were performed. Semmelweis instituted a simple routine of hand-washing with water solutions of chloride of lime, which promptly reduced mortality to 1.27%. Unfortunately, his ideas were met with fierce opposition by the conservative medical community in Vienna, who subjected him to laughter and ridicule. In disgust, Semmelweis left Austria for Budapest. There he became head of the obstetrical division of St. Rochus Hospital, where puerperal fever mortality rates were reduced to below 1% after Semmelweis introduced chlorinated water disinfection.

The man who elucidated the true nature of infection, founded the science of bacteriology, and paved the way for Lister and the antiseptic system in surgery was *Louis Pasteur* (1822-1895). Pasteur was led to his great discoveries regarding bacteria and other microorganisms by his investigations into the process of fermentation. He showed conclusively that fermentation was brought about by some external agent entering the wine. He proved by painstaking experiments under rigorously controlled conditions that meat and fluids like blood did not putrefy if they were kept in such a way that all air was excluded from them. By taking samples of air at different levels Pasteur showed that contamination became less with increasing altitude. Then he proved that the contaminating agents were living organisms (bacteria) which were everywhere—in every room, in the air, on every article of clothing, on furniture, on the ground, and on the skin. He showed that putrefaction was caused by the presence of bacteria and that this applied to putrefaction in foods (milk, wine, and meat), urine, and in wounds.

The application of Pasteur's discoveries to surgical practice was the work of *Joseph Lister* (1827-1912), a young English surgeon who had concluded that it must also be bacteria that caused the suppuration, pus and gangrene which plagued the surgical wards of those days. He determined to prevent the access of organisms by killing them in or on the surface of the wound. Pasteur had shown that heat could kill microbes, but it was impossible to apply heat to a wound without burning the patient, so some chemical substance had to be found. After trying various chemical agents he finally selected carbolic acid, and he insisted that everything which touched the wound, the dressings, the instruments and the fingers, should be treated with this *antiseptic*. He even produced an antiseptic atmosphere by means of a carbolic spray. The clinical results of Lister's first antiseptic system in 1865 included 11 compound fracture cases with only one death, a 9% mortality rate, marking a watershed between the primitive and modern eras of surgery.

Throughout the 19th century, the arguments continued as to whether microorganisms seen in a sick patient were merely coincidental with the illness, or resulted from the changes brought about by the illness itself. In 1882 the German microbiologist *Robert Koch* (1843-1910) formulated three famous postulates to guide scientists searching for disease-causing microbes. Koch argued that to prove an organism causes a disease, microbiologists must show that the organism occurs in every case of the disease; that it is never found as a harmless parasite associated with another disease; and that once the organism is isolated from the body and grown in laboratory culture, it can be introduced into a new host and produce the disease again. (An oft-stated fourth postulate, that the microbe must be isolated again from the second host, was not part of Koch's original formulation.) Koch and his pupils discovered specific bacillary causes for various diseases, including anthrax, cholera, tuberculosis, gonorrhea, diphtheria, leprosy, typhoid, trypanosomiasis, and malaria.

The theory that specific germs could cause specific disease remained contentious until the beginning of the 20th century. Many scientists of great repute rejected Koch's conclusions, with one scientist confidently asserting that "no microbe found in the living blood of any animal was pathogenic." In one celebrated case, *Max von Pettenkofer* of Bavaria (1818-1901), a distinguished 19th century experimental hygienist, induced Koch to send him a sample of his cholera vibrios culture and then wrote a letter back to Koch in 1892, as follows:

"Herr Doctor Pettenkofer presents his compliments to Herr Doctor Professor Koch and thanks him for the flask containing the so-called cholera vibrios, which he was kind enough to send. Herr Doctor Pettenkofer has now drunk the entire contents and is happy to be able to inform Herr Doctor Professor Koch that he remains in his usual good health."

Apparently Pettenkofer, aged 74 at the time he wrote the letter, survived this cholera exposure quite well, perhaps possessing the high stomach acidity which sometimes neutralizes the bacillus, though he shot himself to death in Munich 9 years later.

Skeptics notwithstanding, microbes were key. Enormous new vistas now lay open, as surgeons could confidently make an incision through intact skin without incurring an extreme risk of wound infection. The next step was to progress beyond killing wound bacteria with chemical antiseptics, to the prevention of bacterial contamination by eliminating bacteria in the operating theater—aseptic surgery. The use of steam sterilization of instruments, dressings and gowns, the wearing of masks, caps and gloves, air filtration and the other rituals of the operating theater of today were introduced over the decades following Lister's efforts.

Anesthesia and antisepsis enabled surgeons to carry out procedures that had formerly been quite beyond them, including long operations inside the head, the abdomen and the pelvis. *Theodor Billroth* of Vienna (1829-1894) resected the esophagus in 1872, parts of the intestines in 1878, and the pyloric end of the stomach in 1881. Billroth also made the first complete excision of the larynx. The first successful repair of a gunshot wound on a major artery was performed in Chicago by *John B. Murphy* (1857-1916) in 1897; *Ludwig Rehn* (1849-1930) of Frankfurt am Main performed the first successful repair of a cardiac injury in Germany in the same year.

Writing in 1874, *Sir John Eric Erichsen* (1818-1896), Professor of Surgery at University College, London, had predicted that "the abdomen, the chest, and the brain would be forever shut from the intrusions of the wise and humane surgeon." By the time of Erichsen's death 22 years later, surgeons had successfully removed from patients the stomach and large parts of the intestines, a whole lung had been excised, and a brain tumour had been extirpated. These operations were not mere feats of surgical showmanship; they saved the lives and restored the health of thousands of human beings.

Cells and tissues

The doctrine of the essential cellular nature of living things was established by 1840. Modern cell theory began in botany. The Jena botanist *Matthias Schleiden* (1804-1881) observed that plants were aggregates of cells, existing as self-reproducing living units. Exploring analogies between animals and plants in structure and growth, Theodor *Schwann* (1810-1882) took up the idea, maintaining that all these phenomena could also be demonstrated in animal structures. Thus living cells were basic to living things, and cells incorporated a nucleus and an outer membrane.

Jacob Henle (1809-85) applied cell biology to man. His three-volume *Handbuch der systematischem Anatomie des Menschen (Handbook of Systematic Human Anatomy)* (1866-1871) addressed the body from an architectural standpoint, describing its macro- and microscopic structure. Henle discovered kidney tubules and was the first to describe the muscular coat of the arteries, the minute anatomy of the eye and various skin structures, earning him a reputation as the Vesalius of histology (the study of tissues). Physiologists stressed organ and tissue function—*Claude Bernard* (1813-1878) emphasized the experimental

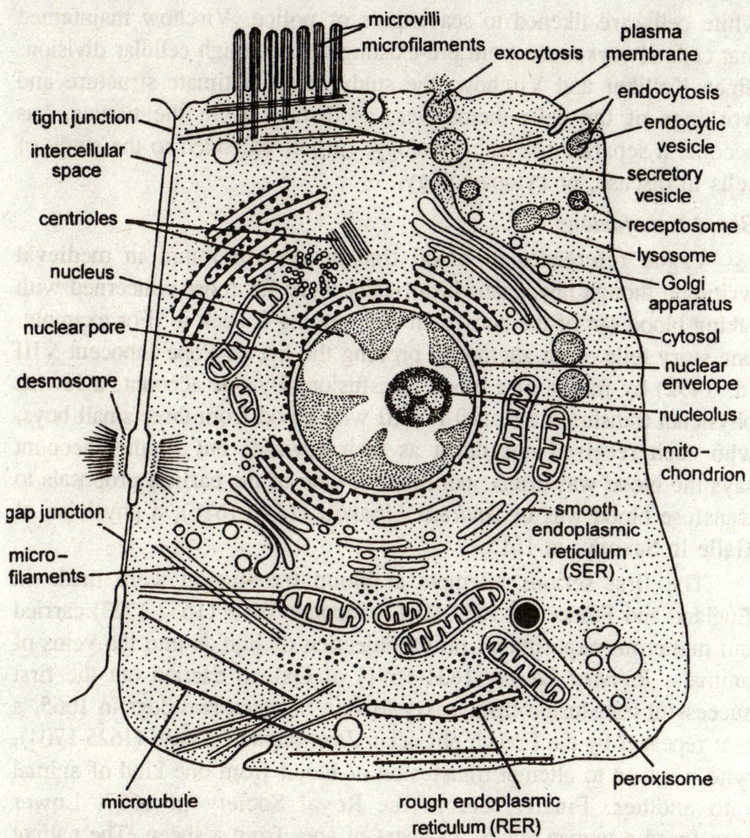

Fig. 1.2. Ultrastructure of a typical animal cell as seen in the electron microscope.

method in establishing biological knowledge and urged that medical practice should be grounded in such knowledge.

Histology was raised to the status of an independent science by the Swiss microanatomist *Albert von Kolliker* (1817-1905), who wrote the first textbook on the subject, *Handbuch der Gewebelehre des Menschen (Handbook of the Tissues of Man)* (1852). The medical implications of cell theory were taken up by *Rudolph Virchow* (1821-1902), who dominated German biomedical research for half a century. Virchow extended the cell concept to diseased tissues; his *Die Cellularpathologie (Cellular Pathology)* (1858) analyzed such tissues from the point of view of cell formation and cell structure. Virchow initiated the idea that the body may be regarded as a "cell state in which every cell is a citizen." Disease is often but civil war, and white cells are likened to scavengers or police. Virchow maintained that cells always arose from pre-existing cells through cellular division. Since Kolliker and Virchow, the study of the intimate structure and workings of the cells themselves, as distinct from the tissues, has become a separate science, cytology, further extended to the study of cells in disease, or cytopathology.

Blood transfusions

Vague references to blood transfusion are found in medieval writings, though physicians historically were far more concerned with taking blood out of the body than with putting it back in. For example, one story tells of an attempt to prolong the life of Pope Innocent VIII (d. 1492) by means of a blood transfusion. By one account, a Jewish physician transfused the aged Pontiff with blood from three small boys, who each received one ducat as their reward; but another account says the blood was drunk, not infused. One of the earliest proposals to transfuse blood was by *Andreas Libavius* (1540-1616), a physician of Halle in Saxony, in 1615.

The first serious attempts at blood transfusion were made in England and France. In 1657 *Sir Christopher Wren* (1632-1723) carried out numerous experiments on the injection of liquids into the veins of animals. *Richard Lower* (1631-1691) of Oxford carried out the first successful transfusion from artery to vein between two dogs in 1665, a feat repeated by the French physician *Jean-Baptiste Denys* (1625-1704), who went on to attempt transfusion of blood from one kind of animal into another. Finally, before the Royal Society in 1667, Lower transfused a human youth, 15 years of age, from a sheep. The patient was greatly improved and the only ill effect was a feeling of great heat along his arm; subsequent patients were not so lucky. Blood transfusions from lambs and calves to humans continued to be tried in the mid 17th century, but were not very successful, and in 1670 were finally forbidden by law in England.

The first known transfusion of human blood into an already moribund person was attempted by *James Blundell* in 1818, and on several other occasions during the 1820s, with poor results. Transfusion was carried out on a small scale during the American Civil War, but technical difficulties connected with premature clotting and the occurrence of accidents arising from the use of incompatible blood prevented the rapid acceptance of the process. The cause of many of the untoward effects of blood transfusion was finally explained in 1901 when the presence of agglutinins and iso-agglutinins in the blood was

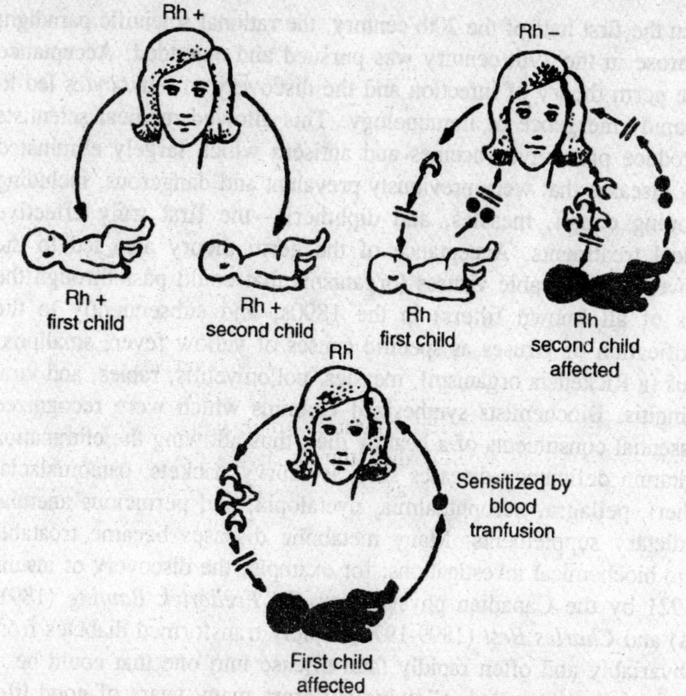

Fig. 1.3. Blood rhesus factor.

demonstrated by *Karl Landsteiner* (1868-1943) in Vienna, winning him the Nobel Prize in 1930; in 1907, the four main blood groups were determined by *Jan Jansky* (1873-1921) of Prague. *Rhesus factor* was later identified by two American scientists, *Philip Levine* (1900-1987) and *Rufus Stetson* (1886-1967). These advances were of fundamental importance and it became possible in the 20th century to eliminate most of the fatalities due to incompatibility, allowing blood transfusions to be practiced reliably and with uniformly good results.

20th century medicine

The 20th century saw more discoveries and advances in medical science than all previous centuries combined. In this period, medicine became more powerful than ever before as scientists gained knowledge of matters and processes of illness that, at the beginning of the century, were still unknown or mysterious. Unlike the mere palliatives of earlier eras, 20th century physicians could actually cure some diseases, reverse some physical traumas, and save many lives that could not be saved before.

In the first half of the 20th century, the rational scientific paradigm that arose in the 19th century was pursued and extended. Acceptance of the germ theory of infection and the discovery of leukocytes led to the rapid emergence of immunology. This allowed medical scientists to produce protective vaccines and antisera which largely eliminated many diseases that were previously prevalent and dangerous, including whooping cough, measles, and diphtheria—the first truly effective medical treatments. Acceptance of the germ theory also led to the discovery of filterable viruses (organisms that could pass through the pores of all known filters) in the 1890s, and subsequently to the identification of viruses as specific causes of yellow fever, smallpox, typhus (a Rickettsia organism), measles, poliomyelitis, rabies, and viral meningitis. Biochemists synthesized vitamins which were recognized as essential constituents of a healthy diet, thus allowing the elimination of vitamin deficiency diseases such as scurvy, rickets, osteomalacia, beriberi, pellagra, xerophthalmia, nyctalopia, and pernicious anemia, via dietary supplements. Many metabolic diseases became treatable due to biochemical investigations; for example, the discovery of insulin in 1921 by the Canadian physiologists Sir *Frederick Banting* (1891-1941) and *Charles Best* (1899-1978) rapidly transformed diabetes from an invariably and often rapidly fatal disease into one that could be at least partially controlled, allowing sufferers many years of good life.

Blood group specification made transfusions convenient, facilitating dramatic advances in many branches of medicine,especially surgery. The first report of a successful autotransplant of a kidney into the neck of a dog was performed by *Emerich Ullmann* (1861-1937) in 1902. The first successful kidney graft between identical twins was performed in 1954 by *Joseph E. Murray* (b. 1919), who received a Nobel Prize for his work. The first heart transplantation in man was achieved by *Christiaan Barnard* (b. 1922) in 1967. Results were poor at first, but the use of new anti-rejection techniques such as the drug cyclosporin in the late 1970s, aided by advances in immunology, greatly improved the success of this and other organ transplants. By 1987, a total of ~7000 human hearts had been transplanted. At the close of the 20th century, heart, lung, heart-lung, and liver transplants were standard procedures, while grafts of small intestine and pancreas (the latter for refractory diabetes) were under active clinical investigation.

Two pivotal events transformed scientific medicine from a merely rational basis to a molecular basis, thus laying the groundwork for 21st century nanomedicine. The first pivotal event was the drug revolution, among which the most useful and spectacular were the antibiotics introduced between 1935-1945 and widely used ever since. *Antibiotics* are significant because they actively interfere with microbial metabolism and growth at the molecular level. The first antibiotic drugs were the *sulphonamides*, in 1935. Then penicillin became available in the 1940s, initially in very small quantities, then mass-produced by Pfizer during and after World War II, as a result of the research work of *Alexander Fleming* (1888-1955), *Howard Florey* (1898-1968) and *Ernst Chain* (1906-1979). In 1943, *Selman A. Waksman* (1888-1973) discovered streptomycin, the first effective anti-tuberculosis drug, for which he received the 1952 Nobel Prize. For the first time, physicians had true cures for many diseases,especially the most common bacterial diseases. Antifungal, anti-parasitic, and antiviral drugs of more limited effectiveness soon followed.

The 20th century also produced drugs that altered mood and levels of consciousness. *Barbiturates* were first introduced in 1903 (e.g. barbitone or Veronal), followed by phenobarbitone (Luminal) in 1912 and *Evipan*, the barbiturate anesthetic, in 1932. By mid-century these highly addictive drugs began to be replaced by the somewhat less-addictive *benzodiazepines*, including *Valium* and *Librium*. Tranquilizers, largely the phenothiazines such as *chlorpromazine* (*Thorazine*) and antimanics such as *lithium carbonate*, came to be widely used in psychiatry as effective medications for major mental illnesses including schizophrenia and manic depression.

The second pivotal event was the genetics revolution, starting with the discovery in 1953 of the information-carrying double-helix structure of DNA by *Francis Crick* (b. 1916) and *John B. Watson* (b. 1928), followed in the 1980s by the ability to chemically read the genetic code, isolate specific genes and clone them for further study. In the mid-1980s, the Human Genome Project was launched, with the objective of fully sequencing every gene in the human genome. The first phase of this project neared completion as the 20th century drew to a close.

Molecular biology became the premier scientific discipline of the latter 20th century. By 1998, the compositions of organs, tissues, cells, organelles, and membranes had been defined, and the biosynthesis and catabolism of hundreds of compounds had been elucidated. The regulation of body processes was described at progressively finer levels in biochemical language. Many pharmacologic agents were finally understood in terms of specific molecular loci and mechanisms of action. Advances were particularly rapid inimmunology, virology, cellular biology, peptide research, and structural biology. A beginning was made in explaining human behaviour in mechanistic terms, as more and more chemical mediators and pharmacologic modifiers were discovered. In biology, these disciplines developed porous boundaries with related disciplines such as physiology, pharmacology, neurosciences, biochemistry and biophysics. All entered a phase of confluence, employing the common language of chemistry.

Thus the late 20th century is best regarded as the molecular age of basic biological science. The molecular influence pervades all the traditional disciplines underlying clinical medicine. As of February 1999, one source listed exactly 1446 genetic disorders, most of which could be linked to a specific human chromosome; another source stated that ~4000 genetic disorders were known in 1998. There were more than 575 known abnormal human hemoglobins, and for each of these the precise structural defect in the DNA of the mutant gene could be defined. Knowledge of membrane, cytoplasmic, and nuclear receptors for hormones and drugs was exploding, with old as well as new diseases being defined in terms of receptor abnormalities—for example, type II hyper-cholesterolemia and nephrogenic diabetes insipidus. Recognition of opiate receptors led to the discovery of endogenous peptides (endorphins) with analgesic activity. Their localization promised further understanding of the limbic system, affective states, and addictions. Defects in a subcellular organelle, the peroxisome, were known to be responsible for a growing list of important genetic afflictions such as Refsum disease, Zellweger syndrome, and X-linked adrenoleukodystrophy. The genetic defect responsible for Huntington's disease was discovered, and in the 1990s the genetic defects responsible for many other important neurological disorders were becoming known, including forms of Alzheimer's disease, Charcot-Marie-Tooth disease, and common blinding retinal degenerative disorders such as retinitis pigmentosa, Leber's hereditary optic neuropathy, Norrie's disease, and choroideremia.

DNA sequencing techniques and restriction endonucleases permitted precise identification of the exact structural alteration of the gene in an increasing number of hereditary diseases. For example, Burkitt's lymphoma is characterized by a translocation of the distal end of the long arm of chromosome 8 to loci on chromosomes 14, 22, or 2. Gene therapy—both pharmacologic modification of specific gene action and physical replacement of damaged genetic segments—became possible in experimental systems. Complete maps of the entire genomes of 18 microbial species had been compiled and published by the end of 1998, with more than 60 others in progress. In 1992, wrote the conservative *Cecil Textbook of Medicine*: "The expansion of the knowledge bank of the past quarter century justifies great optimism for the eventual control and cure of major diseases and the possible elimination of premature death from illness."

Global changes in progressive aging dysfunction were shown to be strongly related to declining secretion of growth hormone by Rudman in 1990. Shortly thereafter, anti-aging medicine was recognized as a distinct discipline and was promoted by the American Academy of Anti-Aging Medicine (A^4M), a group that claimed >6000 physician and scientist members worldwide by 1998 and had held numerous conferences. In a book on the subject, *Ronald Klatz*, A^4M's president, first comprehensively documented the implications of Rudman's work.

While biological science was vaulting forward into the molecular realm at a blistering pace, biomedical engineering lagged considerably behind, though many remarkable successes had been achieved since mid-century. Radiology expanded with sophisticated radiotherapy, ultrasound, scanning and imaging techniques (e.g., CAT, PET, NMR), with submillimeter resolution in living tissues. Surgical instruments became less damaging and less invasive; minimally invasive techniques used OK (orifice and keyhole) surgery and surgery performed under the "eye" of a scan. Fetuses could be screened for abnormality and surgery then performed upon them inside the womb, or they could be removed temporarily from the womb and then returned to continue gestating, after surgery. Prospective parents with fertility difficulties could make use of a wide range of therapies including *in vitro* fertilization. Implantable pacemakers, defibrillators, and ventricular assist devices were commonplace, and in 1998 artificial wearable/implantable and full/partial replacements for lungs, heart, kidney, liver, and pancreas were either available, in clinical trials, or under development. Electron microscopes with 200,000X magnifications allowed many details of internal cellular and viral structures to be resolved as early as 1946; by 1998, atomic-force (AFM) and scanning-tunneling microscopy (STM) permitted direct tactile examination of individual biomolecules in fixed cells.

As the 20th century drew to a close, a few preliminary efforts had been made to apply the molecular approach to medical diagnostic and clinical tools. Biotechnology was one avenue being pursued, with the rational design of artificial enzymes and specified-ligand binding sites having already been achieved in certain limited cases, and the beginnings of gene therapy as noted earlier. Carbon fullerenes had been used to create a water-soluble inhibitor of HIV protease, and had other biological applications. AFM-based force-amplified biological sensors could detect defined biological species such as cells, proteins, toxins, and DNA at concentrations as low as 10^{-18} M (~ $1/mm^3$), and automated laboratory systems for sorting and handling individual cells and viruses were commonplace. Biotech companies such as Physiome Sciences of Princeton NJ had developed three-dimensional computer models of the heart and other organs. Physiome's heart model was based on detailed molecular, biochemical, cellular and anatomical information, including submodels of all the different cell types found in the heart embodying knowledge of the function of each cell type in healthy and diseased hearts, and information on gene function and the causes and effects of congestive heart failure, arrhythmias and heart attacks. Other cell biochemistry simulators included E-CELL and the Virtual Cell. There was also much progress and interest in nanostructure analysis and nanomaterials fabrication for medical and biological purposes, with research groups emerging at major universities such as the Cornell Nanofabrication Facility, theUniversity of Michigan Center for Biologic Nanotechnology, the Rice University Center for Nanoscale Science and Technology, the CalTech Materials and Process Simulation Center, the Washington University Nanotechnology Center (St. Louis, MO), the USC Laboratory for Molecular Robotics, the UCLA Exotic Materials Center, the Institute for Molecular Medicine of the University of Oxford, and at many biotechnology-oriented corporations such as Nanogen and Affymetrix.

However, the greatest medical revolution of all awaits the ability to engineer and fabricate whole devices and systems at the molecular scale. Along this course lies nanotechnology and molecular manufacturing, a deep well from which nanomedicine will inevitably spring.

21st century medicine

It is always somewhat presumptuous to attempt to predict the future, but in this case we are on solid ground because most of the prerequisite historical processes are already in motion and all of them appear to be clearly pointing in the same direction.

Medical historian *Roy Porter* notes that the 19th century saw the establishment of what we think of as scientific medicine. From about the middle of that century the textbooks and the attitudes they reveal are recognizable as not being very different from modern ones. Before that, medical books were clearly written to address a different mind-set.

But human health is fundamentally biological, and biology is fundamentally molecular. As a result, throughout the 20th century scientific medicine began its transformation from a merely rational basis to a fully molecular basis. First, antibiotics that interfered with pathogens at the molecular level were introduced. Next, the ongoing revolutions in genomics, proteomics and bioinformatics provided detailed and precise knowledge of the workings of the human body at the molecular level. Our understanding of life advanced from organs, to tissues, to cells, and finally to molecules, in the 20th century. By the early 21st century, the entire human genome will be mapped. This map will inferentially incorporate a complete catalog of all human proteins, lipids, carbohydrates, nucleoproteins and other molecules, including full sequence, structure, and much functional information. Only some systemic functional knowledge, particularly neurological, may still be lacking by that time.

This deep molecular familiarity with the human body, along with simultaneous nanotechnological engineering advances, will set the stage for a shift from today's molecular scientific medicine in which fundamental new discoveries are constantly being made, to a molecular technologic medicine in which the molecular basis of life, by then well-known, is manipulated to produce specific desired results. The comprehensive knowledge of human molecular structure so painstakingly acquired during the 20th and early 21st centuries will be used in the 21st century to design medically-active microscopic machines. These machines, rather than being tasked primarily with voyages of pure discovery, will instead most often be sent on missions of cellular inspection, repair, and reconstruction. In the coming century, the principal focus will shift from medical science to medical engineering. Nanomedicine will involve designing and building a vast proliferation of incredibly efficacious molecular devices, and then deploying these devices in patients to establish and maintain a continuous state of human healthiness.

The very earliest nanotechnology-based biomedical systems may be used to help resolve many difficult scientific questions that remain. They may also be employed to assist in the brute-force analysis of the most difficult three-dimensional structures among the 100,000-odd proteins of which the human body is comprised, or to help ascertain the precise function of each such protein. But much of this effort should be complete within the next 20-30 years because the reference human body has a finite parts list, and these parts are already being sequenced, geometered and archived at an ever-increasing pace. Once these parts are known, then the reference human being as a biological system is at least physically specified to completeness at the molecular level. Thereafter, nanotechnology-based discovery will consist principally of examining a particular sick or injured patient to determine how he or she deviates from molecular reference structures, with the physician then interpreting these deviations in light of their possible contribution to, or detraction from, the general health and the explicit preferences of the patient.

In brief, nanomedicine will employ molecular machine systems to address medical problems, and will use molecular knowledge to maintain human health at the molecular scale.

Volitional Normative Model of Disease

What, exactly, is "medicine"? Dictionaries give several definitions, ranging from the very restrictive to the most general, as follows: "a drug or remedy"; "any substance used for treating disease"; "any drug or other substance used in treating disease, healing, or relieving pain"; "in a restricted sense, that branch of the healing art dealing with internal diseases"; "treatment of disease by medical, as distinguished from surgical, treatment"; "the branch of this science and art that makes use of drugs, diet, etc., as distinguished especially from surgery and obstetrics"; "the study and treatment of general diseases or those affecting the internal parts of the body"; "the science of treating disease, the healing art"; "the art and science of preventing or curing disease"; "the act of maintenance of health, and prevention and treatment of disease and illness"; "the department of knowledge and practice dealing with disease and its treatment"; or, most generally, "the science and art of diagnosing, treating, curing, and preventing disease, relieving pain, and improving and preserving health". In this book, we shall adopt the latter, maximally-inclusive, definition of "medicine".

The contemporary physician might at first be inclined to relegate molecular approaches to some minor subfield, perhaps "nanoanalytics," "nanogenomics," or "nanotherapeutics." This would be a serious mistake, because the application of molecular approaches to health care will significantly impact virtually every category of laboratory and clinical practice across the board. Thus we are led to the broadest possible conception of nanomedicine as "the science and technology of diagnosing, treating, and preventing disease and traumatic injury, of relieving pain, and of preserving and improving human health, using molecular tools and molecular knowledge of the human body."

This brings us to the question of "disease," a complex term whose meaning is still hotly debated among medical academics. Illnesses due to microorganisms, or conditions in which the doctor's contribution to the diagnosis was important, were most likely to be called a disease, but if the cause was a known physical or chemical agent the condition was less likely to be regarded as disease; general practitioners also had the broadest definition of disease. No less than eight different types of disease concepts are held by at least some people currently engaging in clinical reasoning and practice, including:

Disease nominalism

A disease is whatever physicians say is a disease. This approach avoids understanding and forestalls inquiry, rather than furthering it.

Disease relativism

A disease is identified or labeled in accordance with explicit or implicit social norms and values at a particular time. In 19th century Japan, for example, armpit odor was considered a disease and its treatment constituted a medical specialty. Similarly, 19th-century Western culture regarded masturbation as a disease, and in the 18th century, some conveniently identified a disease called drapetomania, the"abnormally strong and irrational desire of a slave to be free." Various non-Western cultures having widespread parasitic infection may consider the lack of infection to be abnormal, thus not regarding those who are infected as suffering from disease.

Sociocultural disease

Societies may possess a concept ofdisease that differs from the concepts of other societies, but the concept may also differ from that held by medical practitioners within the society itself. For instance, hypercholesterolemia is regarded as a disease condition by doctors but not by the lay public; medical treatment may be justified, but persons with hypercholesterolemia may not seek treatment, even when told of the condition. Conversely, there may be sociocultural pressure to recognize a particular condition as a disease requiring treatment, such as alcoholism and gambling.

Statistical disease

A condition is a disease when it is abnormal, where abnormal is defined as a specific deviation from a statistically-defined norm. This approach has many flaws. For example, a statistical concept makes it impossible to regard an entire population as having a disease. Thus tooth decay, which is virtually universal in humans, is not abnormal; those lacking it are abnormal, thus are "diseased" by this definition. More reasonably, a future highly-aseptic society might regard bacterium-infested 20th century humans (who contain in their bodies more foreign microbes than native cells) as massively infected. Another flaw is that many statistical measurables such as body temperature and blood pressure are continuous variables with bell-shaped distributions, so cutoff thresholds between "normal" and "abnormal" seem highly arbitrary.

Infectious agency

Disease is caused by a microbial infectious agent. Besides excluding systemic failures of bodily systems, this view is unsatisfactory because the same agent can produce very different illnesses. For instance, infection with hemolytic *Streptococcus* can produce diseases as different as erysipelas and puerperal fever, and Epstein-Barr virus is implicated in diseases as varied as Burkitt's lymphoma, glandular fever, and naso-pharyngeal carcinoma.

Disease realism

Diseases have a real, substantial existence regardless of social norms and values, and exist independent of whether they are discovered, named, recognized, classified, or diagnosed. Diseases are not inventions and may be identified with the operations of biological systems, providing a reductionistic account of diseases in terms of system components and subprocesses, even down to the molecular level. One major problem with this view is that theories may change over time —almost every 19th century scientific theory was either rejected or highly modified in the 20th century. If the identification of disease is connected with theories, then a change in theories may alter what is viewed as a disease. For example, the 19th century obsession with constipation was reflected in the disease labelled "autointoxication," in which the contents of the large bowel were believed to poison the body. Consequently much unnecessary attention was paid to laxatives and purgatives and, when surgery of the abdomen became possible toward the end of the century, operations to remove the colon became fashionable in both England and America.

Disease idealism

Disease is the lack of health, where health is characterized as the optimum functioning of biological systems. Every real system inevitably falls short of the optimum in its actual functioning. But by comparing large numbers of systems, we can formulate standards that a particular system ought to satisfy, in order to be the best of its kind. Thus "health" becomes a kind of Platonic ideal that real organisms approximate, and everyone is a less than perfect physical specimen. Since we are all flawed to some extent, disease is a matter of degree, a more or less extreme variation from the normative ideal of perfect functioning. This could be combined with the statistical approach, thus characterizing disease as a statistical variation from the ideal. But this view, like the statistical, suffers from arbitrary thresholds that must be drawn to qualify a measurable function as representing a diseased condition.

Functional failure

Organisms and the cells that constitute them are complex organized systems that display phenomena (e.g. homeostasis) resulting from acting upon a program of information. Programs acquired and developed during evolution, encoded in DNA, control the processes of the system. Through biomedical research, we write out the program of a process as an explicit set (or network) of instructions. There are completely self-contained "closed" genetic programs, and there are "open" genetic programs that require an interaction between the programmed system and the environment, e.g. learning or conditioning. Normal functioning is thus the operation of biologically programmed processes, e.g. natural functioning, and disease may be characterized as the failure of normal functioning. One difficulty with this view is that it enshrines the natural as the benchmark of health, but it is difficult to regard as diseased a natural brunette who has dyed her hair blonde in contravention of the natural program, and it is quite reasonable to regard the mere possession of an appendix as a disease condition, even though the natural program operates so as to perpetuate this troublesome organ. A second weakness of this view is that disease is still defined against population norms of functionality, ignoring individual differences. As a perhaps overly simplistic example, 65% of all patients employ a cisterna chyli in their lower thoracic lymph duct, while 35% have no cisterna chyli—which group has a healthy natural program, and which group is "diseased"?

A ninth view of disease is also proposed, a new alternative which seems most suitable for the nanomedical paradigm, called the "volitional normative" model of disease. As in the "*disease idealism*" view, the volitional normative model accepts the premise that health is the optimal functioning of biological systems. Like the "functional failure" view, the volitional normative model assumes that optimal functioning involves the operation of biologically programmed processes.

However, two important distinctions from these previous views must be made. First, in the volitional normative model, normal functioning is defined as the optimal operation of biologically programmed processes as reflected in the patient's own individual genetic instructions, rather than of those processes which might be reflected in a generalized population average or "Platonic ideal" of such instructions; the relative function of other members of the human population is no longer determinative. Second, physical condition is regarded as a volitional state, in which the patient's desires are a crucial element in the definition of health. This is a continuation of the current trend in which patients frequently see themselves as active partners in their own care.

In the volitional normative model, disease is characterized not just as the failure of "optimal" functioning, but rather as the failure of either (a) "optimal" functioning or (b) "desired" functioning. Thus disease may result from:

1. A failure to correctly specify desired bodily function (specification error by the patient);
2. A flawed biological program design that doesn't meet the specifications (programming design error);
3. Flawed execution of the biological program (execution error);
4. External interference by disease agents with the design or execution of the biological program (exogenous error); or
5. Traumatic injury or accident (structural failure).

In the early years of nanomedicine, volitional physical states will customarily reflect "default" values which may differ only insignificantly from the patient's original or natural biological programming. With a more mature nanomedicine, the patient may gain the ability to substitute alternative natural programs for many of his original natural programs. For example, the genes responsible for appendix morphology or for sickle cell expression might be replaced with genes that encode other phenotypes, such as the phenotype of an appendix-free cecum or a phenotype for statistically typical human erythrocytes. Many persons will go further, electing an artificial genetic structure which, say, eliminates age-related diminution of the secretion of human growth hormone and other essential endocrines. (The graduated secretion of powerful proteolytic enzymes, perhaps targeted for gene-expression in appropriate organs, may reverse and control the accumulation of highly crosslinked collagenaceous debris; by 1998, many members of the mainstream medical community were already starting to regard aging as a treatable condition.) On the other hand, a congenitally blind patient might desire, for whatever personal reasons, to retain his blindness. Hence his genetic programs that result in the blindness phenotype would not, for him, constitute "disease" as long as he fully understands the options and outcomes that are available to him. (Retaining his blindness while lacking such understanding might constitute a specification error, and such a patient might then be considered "diseased.") Whether the broad pool of volitional human phenotypes will tend to converge or diverge is unknown, although the most likely outcome is probably a population distribution (of human biological programs) with a tall, narrow central peak (e.g., a smaller standard deviation) but with longer tails (e.g., exhibiting a small number of more extreme outliers).

One minor flaw in the volitional normative model of disease is that it relies upon the ability of patients to make fully informed decisions concerning their own physical state. The model crucially involves desires and beliefs, which can be irrational, especially during mental illness, and people normally vary in their ability to acquire and digest information. Patients also may be unconscious or too young, whereupon default standards might be substituted in some cases.

Nevertheless, the volitional normative view of disease appears most appropriate for nanomedicine because it recognizes that the era of molecular control of biology could bring considerable molecular diversity among the human population. Conditions representing a diseased state must of necessity become more idiosyncratic, and may progressively vary as personal preferences evolve over time. Some patients will be more venturesome than others— "to each his own." As an imperfect analogy, consider a group of individuals who each take their automobile to a mechanic. One driver insists on having the carburetion and timing adjusted for maximum performance (the "racer"); another driver prefers optimum gas mileage (the "cheapskate"); still another prefers minimizing tailpipe emissions (the "environmentalist"); and yet another requires only that the engine be painted blue (the "aesthete"). In like manner, different people will choose different personal specifications. One can only hope that the physician will never become a mere mechanic even in an era of near-perfect human structural and functional information; an automobile conveys a body, but the human body conveys the soul. Agrees theorist Guttentag: "The physician-patient relationship is ontologically different from that of a maintenance engineer to a machine or a veterinarian to an animal."

Treatment Methodology

The availability of advanced nanomedical instrumentalities should not significantly alter the classical medical treatment methodology, although the patient experiences and outcomes will be greatly improved. Treatment in the nanomedical era will become faster and more accurate, efficient, and effective. In clinical practice, patient treatment customarily includes up to six distinguishable phases: examination, diagnosis, prognosis, treatment, validation, and prophylaxis. Let us consider each of these, in turn.

Examination

The first step in any treatment process is the examination of the patient, including the individual's medical history, personal functional and structural baseline, and current complaints. In classical medicine, interview and observation have long been the cornerstone of examination. In ancient times this was limited to obvious manifestations and simple constellations of observables, such as the Hippocratic facies, the four signs of inflammation noted by Celsus, or pulse rate and fever. Clinicians recognize that the traditional taking and interpreting of oral medical histories from new patients is a subtle and complex art, although some aspects of this process might be automated using voice recognition and text preinterpretation software, somewhat easing the physician's burden.

Advancing technology has also brought a plethora of tests that contribute to accurate diagnosis, including auscultation, microscopy and clinical bacteriology in the 19th century, and radiological scanning, clinical biochemistry, genetic testing, and minimally invasive exploratory surgery in the 20th century.

In the 21st century, new tools for nanomedical testing and observation will include clinical *in vivo* cytography; real-time whole-body microbiotic surveys; immediate access to laboratory-quality data on the patient (e.g. blood tests such as blood counts, dissolved gases and solutes, vitamin and ion assays); physiological function and challenge tests; tissue composition including direct organelle counts in specified tissue populations; quantitative flowcharts of *in cyto* secondary messenger molecules, extracellular hormones and neuropeptides; per-compartment cytoglucose inventories; and so forth. Before a proper diagnosis can be made, the physician must also establish the patient's personal functional and structural baseline against which any deviations can be noted and corrected, in keeping with the volitional normative model of disease.

The capabilities of nanomedial testing are explored at length. By way of introduction, it is instructive to think about a trivial class of test procedures that might be used to diagnose a simple infectious disease at several different levels of technological competence. Let us consider a patient who presents with signs and symptoms that are nonspecific in nature but which suggest an infectious process— e.g. nasal congestion, mild fever, discomfort and cough. The initial signs are due in part to the body's inflammatory response and in part to the infectious agent itself. The diagnostic goal is to identify the infectious agent.

In the late 20th century, the usual procedure would be to culture a sample taken from the patient, in the microbiology laboratory, using various broths, petri plates, and biochemical tests. Some infectious agents are easy to demonstrate. Beta hemolytic streptococci from a throat swab will grow overnight on a blood agar plate, and colony counts for *E. coli* in a urine sample are available in 24 hours. A throat culture that the lab reports as a mixed culture causes no excitement, and a single isolate of *Staphylococcus epidermidis* in a blood culture is usually regarded as a skin contaminant.

Moving up to a higher level of technological competence, biotechnologists describe an ideal diagnostic scenario which takes a molecular approach to the diagnosis of infectious disease using recombinant DNA technology. This approach was not yet possible in 1996 when first suggested, but was regarded as a reasonable and likely future application of biotechnology in the early 21st century given rapid progress in single-molecule DNA assay techniques:

"A patient presents in the clinic with mild fever, nasal congestion, discomfort, and cough. A swab of his throat is taken. Instead of culture to identify abnormal microorganisms by their pattern of growth, the sample is analyzed by recombinant DNA techniques. The cotton throat swab is mixed with a cocktail of DNA probes. Enzymes that digest and release the DNA from both host cells and invading bacteria make the DNA in the sample immediately available for hybridization to the probes. The swab is swirled in the liquid mix of the prepackaged test kit for 1 minute. The liquid is then poured through a column that separates hybridized DNA molecules (bacterial target DNA sequences bound to probe DNA) from all other debris [taking several minutes]. A chemi-luminescence detection system for the probes shows two of several possible colours indicating mixed infection. The diagnostic result, available in 10 minutes, indicates a Rhinovirus of a strain known to be epidemic in the geographic area. A significant superinfection with a penicillin-resistant streptococcus is also identified. With a definitive diagnosis, the patient is started on the appropriate antibiotic."

How might nanomedicine handle this test? In the nanomedical era, taking and analyzing microbial samples will be much simpler for the practitioner. Such analysis will be as quick and convenient as the electronic measurement of body temperature using a tympanic thermometer in a late 20th-century clinical office or hospital. The physician faces the patient and pulls from his pocket a lightweight handheld device resembling a pocket calculator. He unsnaps a self-sterilizing cordless pencil-sized probe from the side of the device and inserts the business end of the probe into the patient's opened mouth in the manner of a tongue depressor. The ramifying probe tip contains billions of nanoscale molecular assay receptors mounted on hundreds of self-guiding retractile stalks. Each receptor is sensitive to one of thousands of specific bacterial membrane or viral capsid ligands. An acoustic echolocation transceiver provides gross spatial mapping. The patient says "Ahh," and a few seconds later a three-dimensional colour-coded map of the throat area appears on the display panel that is held in the doctor's hand. A bright spot marks the exact location where the first samples are being taken. Underneath the colour map scrolls a continuously updated microflora count, listing in the leftmost column the names of the ten most numerous microbial and viral species that have been detected, key biochemical marker codes in the middle column, and measured population counts in the right column. The number counts flip up and down a bit as the physician directs probe stalks to various locations in the pharynx to obtain a representative sampling, with special attention to sores or any signs of exudate. After a few more seconds, the data for two of the bacterial species suddenly highlight in red, indicating the distinctive molecular signatures of specific toxins or pathological variants. One of these two species is a known, and unwelcome, hostile pathogen. The diagnosis is completed, the infectious agent is promptly exterminated, and a resurvey with the probe several minutes afterwards reveals no evidence of the pathogen.

Diagnosis

Diagnosis is the determination of the cause and nature of a disease in order to provide a logical basis for treatment and prognosis. Traditionally the diagnostic process begins with a thorough history taken from the patient and a relevant physical examination. Often this sufficed to make a confident diagnosis, but the cause of some illnesses remained uncertain without recourse to additional information such as blood tests or radiological examinations. Nanotechnology-based diagnosis will consist principally of examining the patient to determine how he or she deviates from autogenous reference structures and functions, and then interpreting those deviations as healthy or unhealthy for that patient.

In the 20th century, diagnoses frequently involved a high degree of uncertainty, largely due to the general lack of comprehensive molecular diagnostic tools. Thus diagnosis would be guided by statistical analyses; one branch of decision analysis, called utility analysis, even allowed the patient to participate in the decisionmaking process. When the correct decision is unclear, urges one textbook, it is well to remember time-honored Hippocratic aphorisms such as "first, do no harm" and "common things occur commonly." The eminent Canadian physician Sir William Osler (1849-1919) lamented that "errors of judgement must occur in the practice of an art which consists largely in balancing probabilities." Most doctors would prefer to understand the root cause of medical problems rather than adopt mere statistical approaches.

Nanomedical tools will vastly reduce diagnostic uncertainty. Using nanomedical instrumentalities, doctors will gain access to unprecedented amounts of information about their patients including in-office comprehensive genotyping and real-time whole-body scans for particular bacterial coat markers, tumour cell antigens, mineral deposits, suspected toxins, hormone imbalances of genetic or lifestyle origin, and other specified molecules, producing three-dimensional maps of desired targets with submillimeter spatial resolution. Embedded *in vivo* nanomedical data archives can provide onboard storage of regularly updated self-diagnostic scans, reducing to a minimum the need for symptomatic interview data from patients who may be unconscious, inarticulate, or verbose, who may have limited powers of self-analysis or self-observation and who may have forgotten, suppressed, or amplified descriptions of symptoms. Physicians do not require an exhaustive survey of the entire body of each patient to molecular detail to make a valid diagnosis. In any particular case, it is the function of the trained medical mind to quickly ascertain where and where not to look in molecular detail. But in the nanomedical era, powerful tools will be available to allow the practitioner to examine almost any portion of a patient in as much detail as desired, right down to the molecular level, with results available in seconds or minutes, and at reasonable cost.

Prognosis and treatment

Prognosis is a judgement or forecast, based upon a correct diagnosis, of the future course of a disease or injury, and of the patient's prospects for partial or full recovery. Guttentag identifies prognosis as "the predicted course of the reduced state of the patient's psychosomatic freedom of action," and treatment as "the physician's ability to intervene."

But prognosis is a function of treatment as well as disease. From the post-Hippocratic era through the 18th century, treatments were almost purely empirical and often did more harm than good. During the 19th and early 20th centuries, treatments were scientific but largely homeostatic—the medical intervention was rational but served mainly to assist the body in healing itself. Throughout the remainder of the 20th century, truly curative treatments began to rescue some patients from conditions from which their unaided bodies would not have been able to recover. Although conventional biotechnology will enable some important tissue and cellular replacement treatments by the early 21st century, nanomedicine will enable major reconstructive and restorative procedures at the tissue, cellular and molecular levels and will employ active antibiotic devices. The prognosis will almost always be good, except in cases of severe neural damage and a few other specialized circumstances. Therapeutic treatments will be selected to reverse all pathological effects of disease or injury, with a minimum of pain, discomfort, side-effects, intrusiveness and time, and with a maximum of effectiveness, efficiency, and likelihood of success, though of course some tradeoffs will always exist. Nanomedicine also will excel in the correction of molecular defects of a kind which Nature has no predesigned tools—such as the breakdown and removal of intracellular lipofuscin (for which there appear to be no natural enzymes) and the removal of indigestible waste products which interfere with neuronal axon transport.

We may compare the therapeutic response to a simple infection at several different levels of technological competence. Consider a patient who has been diagnosed with eastern equine encephalitis, a mosquito- or tick-borne arbovirus. In the 20th century, there was no specific treatment for this disease. Care was generally supportive, with the doctor attempting to maintain the patient's heart and lung function while the infection ran its course. The prognosis was poor.

There was a 50%-75% mortality rate with frequent sequelae including seizures and paralysis, especially in children.

Biotechnologists proposed a molecular approach to therapeutics using recombinant DNA technology that was not yet possible in 1996 but was regarded as likely by the early 21st century:

"A patient enters the hospital with high fever and intense headaches. A spinal fluid tap is submitted to the molecular microbiology lab. After screening for several viruses, a species of equine encephalitis virus is identified that is endemic to a location recently visited by the patient. A call to the Centers for Disease Control results in the emergency delivery of a new antiviral agent. Antisense oligonucleotides are injected into the cerebral spinal fluid. These small DNA pieces bind directly to the virus and block its further proliferation. A temporary reservoir giving access to the cerebrospinal fluid is placed and infusion of tnis therapeutic molecular inhibitor of the virus continues for 5 days until signs of encephalitis have passed."

The nanomedical therapy? As before, a nanomedical cure for eastern equine encephalitis may be far simpler, less painful and a great deal quicker. A single therapeutic dose consisting of ~ 0.1 cm^3 of isotonic saline fluid containing ~ 10 billion active micron-size virucidal nanodevices, a 10% volumetric nanodevice suspension, is injected into the cerebrospinal fluid. Each therapeutic nanorobot has chemical sensors that can unambiguously recognize fluidborne or *in cyto* arbovirus particles and, once recognition has occurred, destroy them and also reverse the cellular damage. A nanorobot population of this size should be able to destroy all viral particles and effect needed repairs in at most an hour, after which the devices are programmed and equipped either to eliminate themselves from the body or to be manually exfused (e.g., nanapheresis).

Validation and prophylaxis

A proper therapeutic protocol will include a procedure for follow-up to ensure that the prescribed treatment was correctly executed with good results. This step is often neglected in order to save costs and may be considered unimportant by some practitioners because approximately 80%-90% of all illnesses which take patients to the doctor are self-curing or self-limiting. For example, the common cold, most infectious diseases and many minor injuries are problems that usually will resolve on their own even with no treatment. In these cases the purpose of treatment is not to provide a cure, but rather to speed the healing process, improve comfort, and avoid complications.

Many nanomedical treatments will require supervision and will run quickly to completion, thus follow-up may come back into vogue. Validation may also be viewed as a post-treatment re-diagnosis to ensure that no disease remains present in the patient.

Prophylaxis is the prevention of disease, typically including patient education, immunization programs, amelioration of occupational hazards, and other preventive and public health measures. In a treatment environment that is rich in effective antibacterial instrumentalities, those microbes which survive will evolve to produce only modest or negligible symptoms that are insufficiently annoying to motivate a patient to seek professional therapeutic relief. It is well-known that bacteria can modify their behaviour over time. For example, syphilis had a much more fulminating course in the Middle Ages than it has in the 20th century. Some future strain of the syphilitic microbe might produce negligible symptoms, but we should still insist on its eradication because of its potential to revert to its earlier virulence if allowed to spread unchecked in a more benign form. Preventative procedures may also be needed to discover, diagnose, and treat apparently symptomless diseases, and a variety of molecular-based physiological malfunctions and structural micropathologies may require nanoscale tools in order to detect them.

With medical conditions that require ongoing supervision and adjustment, such as maintaining optimum hormone balance and minimal accumulation of molecular debris (e.g. anti-aging medicine), nanoscale monitoring stations may act as onboard cellular guidance systems, stimulating or suppressing endocrine secretion as necessary to preserve an ideal state of equilibrium. In some cases, direct manufacture of compounds not easily produced by ribosomes or other biological organelles may be required.

Evolution of Bedside Practice

The relationship between physician and patient has been evolving in response to the rapidly changing medical environment. Two of the most important trends are the decline of traditional holism (with a concomitant increase in specialization) and the rise of therapeutic customization in medical practice.

Specialization and holistic medicine

Holism is a philosophy which holds that individuals function as complete units that cannot be reduced merely to the sum of their parts. It is unarguable that a simplistic reductionist view of the patient which ignores the complex interactions among the many cells, tissues, organs, and systems constituting the human body is deeply flawed. For example, an understanding of the molecular basis of the contractile proteins of heart muscle will not alone tell us how heart muscle cells will look or act; knowing only about the parts of something is not sufficient to predict the behaviour of the whole. However, traditional concepts of holistic medicine go well beyond such basic systems philosophy—incorporating requirements for a consideration of all physical, emotional, social, spiritual, environmental and economic needs of the patient.

N. Jewson and others have decried the modern shift away from holistic medicine, which is asserted to have taken place in three historical phases in the West. The initial phase is identified as the practice of "bedside medicine," where wealthy fee-paying clients in the 17th and 18th centuries helped shape their own diagnosis and treatment by medical practitioners in an holistic manner. Aspects of the patient's emotional and spiritual life were seen as central by the practitioner in making a diagnosis, since most physical treatments were only palliative and so the doctor had to focus on nonphysical supportive measures. This frame of reference was progressively replaced during the 19th century with the trend toward "hospital medicine," wherein physicians concentrated on generic classifications of diseases that were manifested in the patient, moving doctors away from the earlier focus on the individual as a whole person. The 20th century saw the development of "laboratory medicine," which moved diagnosis and therapy even further away from the whole patient, "who came to be medically conceived as little more than a depersonalized object, comprised of a complex of cells."

A more charitable view is that physicians have increasingly specialized in treating those physical diseases for which effective treatments may be readily specified, leaving nonphysical and nontreatable issues for psychiatrists, social workers, fitness coaches, priests, lawyers, or other professionals to deal with. In the 21st century this operational specialization may become complete, since nanomedicine phenomenologically regards the human body as an intricately structured machine with trillions of complex interacting parts, with each part (and each subsystem of parts) subject to individual scrutiny, repair, and possibly replacement by artificial technological means. In this new medical cosmology, the concept of the whole patient almost completely dissolves into a data-intensive whirlwind of molecular detail at the cellular, tissue, organ, and systemic levels.

And yet, as in a bygone era, patients once again will help to shape their own diagnosis and treatment at the hands of medical practitioners who begin to apply the volitional normative model of disease in their practices. This may breathe new life into the age-old medical school dictum to "treat the patient as a person" and to "focus on the sick person" rather than exclusively on the body.

The availability of extremely powerful and transforming molecular technologies also argues for a return to the romantic and perhaps quaint concept of a single doctor taking care of a single patient. The potential for interactions among highly potent nanorobotic instrumentalities argues for diagnostic and therapeutic "gatekeeping" by a single trusted practitioner in whom strategic treatment responsibility is vested—in partnership, of course, with the patient.

Customized diagnosis and therapeutics

In science the objective is to understand the individual occurrence by means of a general law; in medical practice, knowledge of what is generally the case does not tell the physician how to treat a particular patient. Thus the problem of how to proceed from the prototypic case to the individual instance remains to be solved in a systematic manner by the practitioner.

The practitioner of "bedside medicine" in the 17th and 18th centuries had few curative and almost no customized tools at his disposal—perhaps a few dozen basic surgical techniques, performed septically, hazardously, and without anesthesia, and a drug/herbal formulary consisting of a few hundred substances. For instance, the *Edinburgh Pharmacopoeia* of 1803 listed only 222 simples while the *London Pharmacopoeia* of 1809 listed fewer than 200 items, most of which had variable, uncertain, minimal, or nonspecific potency. A few more options became available to the 19th century medical practitioner, including complex surgical techniques for specific conditions that could be performed aseptically with anesthesia and a good chance for success, a somewhat broader and more efficacious pharmacopeia, vaccines targeted to several diseases, and an improving diagnostic ability. By the 20th century, the physician could prescribe from among tens of thousands of specific drugs to target specific bacterial, viral, fungal, or parasitic infections; select from among a very precise array of anesthetic agents, chosen to avoid allergic responses in particular patients; perform a wide variety of noninvasive tests and scans for diagnostic purposes to identify very specific conditions; and perform minimally invasive surgeries directed at many arbitrary tissue masses as small as 1 mm^3 in volume. The first glimmerings of personalized genetic therapies also began to appear.

With the arrival of nanomedicine in the 21st century, the treatment paradigm will complete its transition from coarse-grained, one-size-fits-all, slow-acting methods to molecularly-precise, completely customized, speedy and highly-efficacious procedures and instrumentalities. The irregular shotgun pattern of 18th century palliatives will evolve during the early 21st century into a penetrating and perfectly tailored hail of "magic bullets" each targeting an individual cell or group of cells unique to the individual patient. The 19th century herbalist John Ayrton Paris could have been describing the nanomedical future in his popular textbook *Pharmacologia*, 1840 edition, when he wrote:

"If [a physician] prescribes upon truly scientific principles, he will rarely in the course of his practice compose two formulae that shall, in every respect, be perfectly similar, for the plain reason that he will never meet with two cases exactly alike. Now let me ask what constitutes the essential difference between the true physician and his counterfeit between the philosopher and the empiric? Simply this that the latter exhibits the same medicine in every disease, however widely each may differ from the other in its symptoms and character; while the former examines, in the spirit of philosophic analysis, all the existing peculiarities of his patient, and of his discord...and then adapts with a sound discretion and with a correct judgement of his medicinal agents, such means as may best be calculated to control and correct the patient's morbid condition."

Physician-patient relationship

Many other aspects of the physician-patient relationship, especially as this relationship may evolve in the coming era of nanomedicine, are important and worthy of extensive discussion. One such issue is the obligation of both parties in the partnership to tell the truth. The patient as a fellow human being has every right to know the truth about his or her biological condition, but other considerations may enter into the fulfillment of this obligation. Patients have a quite natural anxiety about their own possible death, and it has been claimed that this anxiety implies that no one can be truly objective toward his or her own body. Guttentag observes that "telling an unwelcome truth to the unprepared is as ill-conceived as trying to hide the truth from the prepared."

In the nanomedical era, the sheer number of "truths" that may become available for disclosure will increase enormously even as the terminal prognosis becomes rare. For example, each human being is believed to possess at least 4-10 potentially serious genetic defects; up to 1% of human DNA is of exogenous viral origin, and as much as 10% of the genome consists of transposons, discrete sequences that are positionally mobile among the chromosomes. Should something be done about this, or not? What should the average patient make of the news that his physician has discovered exactly 57 submicron-scale lamellar defects scattered throughout the compact bone of the caudal epiphysis of the patient's right humerus? In the nanomedical era, people will gain the ability to specify their own physical structure to minute detail, but many patients will not be ready, willing, or able to assume responsibility for this knowledge. Thus there is no ideal substitute for the doctor's interpretative abilities and judgements on the patient's behalf as to the personal significance of specific diagnostic information. As the great clinician Thomas Addis observed in another context: "Honesty with patients requires thought and discipline and effort."

Perhaps the single most important aspect of the physician-patient relationship, in any century, is the humanistic quality of the good doctor. The patient seeks a physician who cares about him as a person and will diagnose and prescribe in a sensitive and compassionate manner, accepting some degree of obligation to the patient. Speaking to medical students, J.C. Bennett describes the implicit social contract between doctor and patient that will still apply in the nanomedical era, as it does today:

"To receive medical care, patients must trust their bodies and their very lives to physicians, and so to be in an honest position to give medical care, physicians must earn such radical trust. Mere technical treatment of disease does not suffice. Patients must be able reasonably to believe that their physicians care about them in an extraordinarily personal way. This exchange of care for trust, while not identical to friendship or love, is equally binding. From it develops an interdependence that is far from unwholesome; rather, it potentiates care and promotes healing. Our late twentieth century sophistication and technologic orientation have too often cost us warmth, humor, and humanity, leaving us in social isolation. We do far better as professionals to err on the side of being human with our patients, than to try to play deus ex machina, the god from the machine."

Changing View of the Human Body

How does a patient regard his or her own body, and how might this most intimate of all relationships change in the nanomedical era?

The so-called dualist theory of the human compound, as originally developed by Descartes and widely accepted today by the ordinary person, holds that the human being consists of two separate kinds of thing: the body and the mind or soul. The body acts as a host or receptacle for the mind. The mind, often called "the ghost in the machine," is manifested by the brain, which it uses (via the bodily senses) to acquire and store information about the world and to integrate this with its genetically-driven imperative to live, thus resolving internal

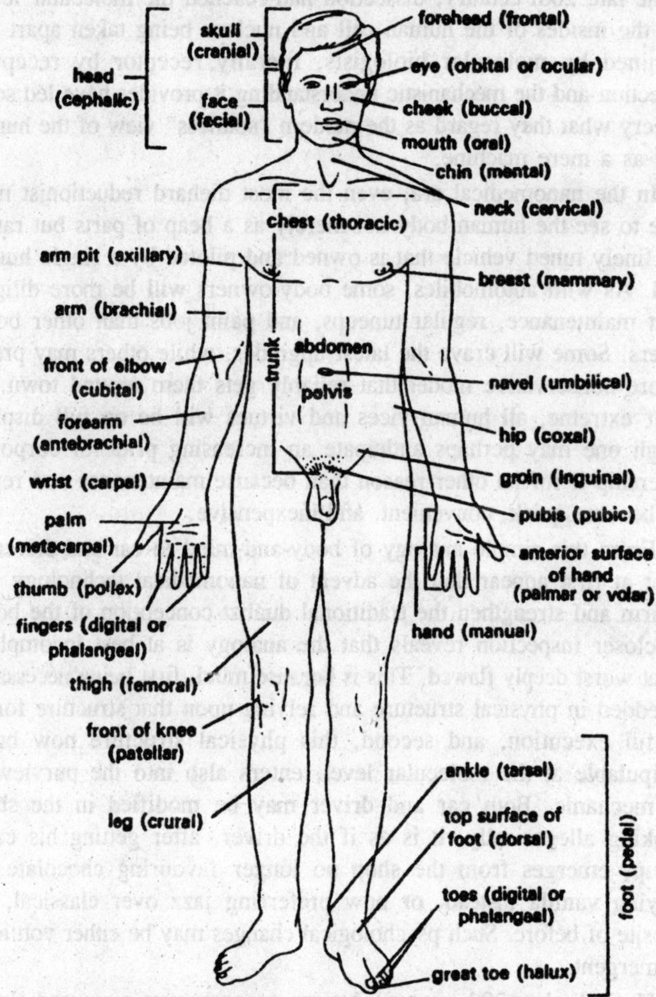

Fig. 1.4. Dorsal view showing the anatomical position.

conflicts among action-choices and expressing these in what we (in our consciousness) experience as decisive action.

Scientific medicine has concentrated primarily on the body. The ancient Roman physician Galen first dissected and vivisected a variety of animals to increase his knowledge of anatomy and physiology, and dissection became increasingly important in the training of physicians and surgeons, and in painting and sculpture, during the Renaissance. By the late 20th century, dissection had reached the molecular level, with the insides of the human cell and nucleus being taken apart and examined by molecular biologists, literally receptor by receptor. Dissection and the mechanistic understanding it provides have led some to decry what they regard as the modern "soulless" view of the human body as a mere machine.

In the nanomedical era, even the most diehard reductionist must come to see the human body not merely as a heap of parts but rather as a finely tuned vehicle that is owned and piloted by a single human mind. As with automobiles, some body-owners will be more diligent about maintenance, regular tuneups, and paint jobs than other body-owners. Some will crave the latest upgrades, while others may prefer a more conservative model that reliably gets them around town. At either extreme, all human vices and virtues will be on full display, though one may perhaps anticipate an increasing pride of corporeal ownership if for no other reason than because maintenance and repair will become quick, convenient, and inexpensive.

From this simple analogy of body-and-mind to car-and-driver, it might at first appear that the advent of nanomedical technology will confirm and strengthen the traditional dualist conception of the body. But closer inspection reveals that the analogy is at best incomplete, and at worst deeply flawed. This is because mind, first being necessarily embedded in physical structure and relying upon that structure for its faithful execution, and second, this physical structure now being manipulable at the molecular level, enters also into the purview of our mechanic. Both car and driver may be modified in the shop. Speaking allegorically, it is as if the driver, after getting his car a tuneup, emerges from the shop no longer favouring chocolate but enjoying vanilla instead, or now preferring jazz over classical, the opposite of before. Such psychological changes may be either volitional or emergent.

Until the late 20th century, human progress was measured almost exclusively in terms of externalities. Food was gathered, then sown, then manufactured. Shelters had no running water, then gained outhouses, then indoor plumbing. Natural lighting and campfires gave way to candles, then oil lamps, then electric illumination. Finger-counting yielded first to the abacus, then the mechanical adding machine, and finally to the digital computer. But throughout all of history, the human body itself has remained largely untouched by progress. We have always regarded our bodies, evolved by natural selection, as fundamentally inviolate and immutable—subject perhaps to various natural or traumatic degradations, but rarely to any significant intrinsic improvement on the timescale of human civilization.

Now we are set to embark upon an era in which our natural physiological equipment may for the first time in history become capable of being altered, improved, augmented, or rendered more comfortable or convenient, due to advances in medical technology. The physical human body may be one of the last bastions of "naturalness". It will also be one of the last elements in our common worldview to be modernized.

Our subjective experience of reality will shift by subtle degrees. For instance, all objective information about our physical surroundings has traditionally arrived in the conscious mind via the various natural senses such as hearing, sight, and smell. In the nanomedical era, machine-mediated sensory modalities may permit direct perception of physical phenomena well removed from our bodies in both time and space, or which are qualitatively or quantitatively inaccessible to our original natural senses. Perception will gradually expand to incorporate nonphysical phenomena including abstract models of mental software, purely artificial constructs of simulated or enhanced realities, and even the mental states of others. Such new perceptions will inevitably alter the way our minds process information.

But the winds of change will sweep deeper still, into our very souls. Like ants oblivious to the collective purpose of their colony, the billions of neurons in the human brain are all busily buzzing, wholly ignorant of the emergent plan. This is the physical, mechanical world of our electrochemical hardware. People also have thoughts, feelings, emotions, and volitions, a higher level in the data processing hierarchy which in turn is equally oblivious of the brain cells. We can happily think while being totally unaware of any help from our neurons. But nanomedicine will give us unprecedented systemic multilevel access to our internal physical and mental states, including real-time operating parameters of our own organs, tissues, and cells, and, if desired, the activities of small groups of (or even individual) neurons. Diverse parts of our selves previously closed to our attention may slowly conjoin and enter our conscious awareness.

Will this access promote an integrated identity or lead to hopeless confusion, or worse? Marvin Minsky, in his collection of essays *The Society of Mind*, persuasively argues that our selves or identities are in fact networks of semi-autonomous neurological "agencies" which sometimes cooperate and sometimes compete with one another. We think of ourselves as singular "persons," but we also experience "conflicting desires" and "differing viewpoints" within our minds that are, in Minsky's view, a direct experience of the multiplicity of our brain's neurostructures. Other models of the human mind suggest that our internal mental states, prospectively transparent via nanomedical augmentation, are diverse and intricate; Julian Jaynes is one of many writers who have drawn attention to profound dichotomies between the two cerebral hemispheres. The component-oriented personality models of Freud (e.g. ego/id/superego), Jung (e.g. archetypes), and Rank (e.g. will/counterwill), and the identification of 4541 distinct personality traits by Allport and Odbert warn us that full access to our brain's architecture could be perilous.

More seriously, most of us suppose that we are endowed with free will. But if choices by free will are simply the resolution of conflicts of neurological subsystems, and we become consciously aware of those subsystems and are able to intervene in their processes, do we run the risk of runaway instabilities at the deepest levels of what we presently call our "minds"? Will we find that these instabilities are profound counterparts to the maladies we currently designate as epilepsy, or psychosomatic illnesses? In any redesigns of our brains which would involve opening doors to, quite literally, the ultrastructure of our thoughts, we could become "naked to ourselves" in ways that we can only vaguely speculate about at present. Along with any other dangers we might encounter, this will raise entirely new issues of the proper role of psychotherapy and the sanctity of personal privacy.

Repairs to the brain may be carefully monitored to ensure quality control and to verify intended results, as already proposed in another context. Major modifications might be strictly regulated, both to prevent abuse by unscrupulous third parties and also to forestall accidental or volitional alterations that could render the patient a significant threat to society. Nanomedical alterations to the brain and other physical systems may give us vastly expanded freedom to be who we choose to be, along with increased responsibility to make wise and informed choices. The ethical and legal aspects of these questions, as well as the scientific and psychological ones, are extremely important and should be thoroughly debated in the years and decades that lie ahead.

Nanomedical Perspective

Nanomedicine and Molecular Nanotechnology

A mature nanomedicine will require the ability to build structures and devices to atomic precision, hence molecular nanotechnology and molecular manufacturing are key enabling technologies for nanomedicine. The prefix "nano-" (from the Greek root nanos, or dwarf) means one-billionth (10^{-9}) of something. The term "nanotechnology" refers most generally to technology on the scale of a billionth of a meter, or a nanometer (a nanometer is ~6 carbon atoms wide). Similarly, the words "nanomachine," "nanorobot," "nanomotor" and "nanocomputer" may refer to complex engineered objects fabricated by positioning matter with molecular control.

Molecular engineering was discussed as an extension of bulk technologies in the 1960s and 1970s by von Hippel, von Foester, and Zingsheim. The phrase "Nano-Technology" was first used in print in 1974 by N. Taniguchi to refer to the increasingly precise machining and finishing of materials, progressing from larger to smaller scales and ultimately to nanoscale tolerances, following in the path of Feynman's proposed "top-down" approach, a schemata which persisted in Taniguchi's thinking throughout the 1980s and 1990s. In 1981, K.E. Drexler described a new "bottom-up" approach involving molecular manipulation and molecular engineering in the context of building molecular machines and molecular devices with atomic precision, a fundamentally different mind-set. Drexler again described molecular technology in 1982 and molecular mechanical devices in 1983, first using the word "nanotechnology" in 1985 and 1986 as synonymous with molecular technology, finally settling upon "molecular nanotechnology" in 1991 and "molecular machine systems" in 1992 to clarify that his concept involved working devices constructed with atomic precision, as distinguished from nanostructured bulk materials, micromachinery, polymeric self-assembly, pure biotechnology, nanolithography, Langmuir-Blodgett thin films, and the like. Drexler's definition—molecular nanotechnology as the three-dimensional positional control of atomic and molecular structure to create materials and devices with molecular precision—is the usage. The first known use of the term "nanomedicine" was in 1991 by Drexler, Peterson, and Pergamit in their popular book *Unbounding the Future*.

Is molecular nanotechnology possible? This question is explicitly addressed, but the bottom line is that molecular nanotechnology violates no physical laws and there exist many possible technical paths leading to useful results. Even by 1985, for example, G. Yamamoto had reported on molecular gears, describing "compounds that exist in conformations which are regarded as static meshed gears with two-toothed and three-toothed wheels and some of them behave as dynamic gears." H. Iwamura prepared a system that formed a chain of beveled molecular gears with ~GHz rotation rates. In 1998, it was generally accepted that molecular nanotechnology would be developed, although there was still some disagreement about how long it would take.

It is often noted that molecular biological systems are themselves nanomachines, constituting an existence proof for molecular nanotechnology. Indeed, biotechnology is one possible implementation pathway for molecular nanotechnology that is being pursued. Such comparisons have a long history. For example, Marcello Malpighi (1628-1694), a professor of medicine in Pisa who discovered the fine structure of the lungs and the capillaries using the microscope, once observed that "our bodies are composed of strings, thread, beams, levers, cloth, flowing fluids, cisterns, ducts, filters, sieves, and other similar mechanisms."

Parallels to living systems as molecular machines were drawn by Changeau, McClaire, Laing, Drexler, and Mitchell, inspiring early thinking in molecular manufacturing. For example, in 1991 Drexler observed:

"Technology-as-we-know-it is a product of industry, of manufacturing and chemical engineering. Industry-as-we-know-it takes things from nature — ore from mountains, trees from forests — and coerces them into forms that someone considers useful. Trees become lumber, then houses. Mountains become rubble, then molten iron, then steel, then cars. Sand becomes a purified gas, then silicon, then chips. And so it goes. Each process is crude, based on cutting, stirring, baking, spraying, etching, grinding, and the like."

"Trees, though, are not crude. To make wood and leaves, they neither cut, stir, bake, spray, etch, nor grind. Instead, they gather solar energy using molecular electronic devices, the photosynthetic reaction centers of chloroplasts. They use that energy to drive molecular machines—active devices with moving parts of precise, molecular structure—which process carbon dioxide and water into oxygen and molecular building blocks. They use other molecular machines to join these molecular building blocks to form roots, trunks, branches, twigs, solar collectors, and more molecular machinery. Every tree makes leaves, and each leaf is more sophisticated than a spacecraft, more finely patterned than the latest chip from Silicon Valley. They do all this without noise, heat, toxic fumes, or human labour, and they consume pollutants as they go. Viewed this way, trees are high technology. Chips and rockets are not."

Contemplating applications of nanotechnology to medicine, Brian Wowk concluded that 20th century physicians were in a predicament similar to that which would be faced by 18th-century engineers trying to maintain a 20th-century automobile—repairs would be crude at best, and breakdowns inevitable:

"Like primitive engineers faced with advanced technology, medicine must 'catch up' with the technology level of the human body before it can become really effective. What is the technology level? Since the human body is basically an extremely complex system of interacting molecules (i.e., a molecular machine), the technology required to truly understand and repair the body is molecular machine technology—nanotechnology. A natural consequence of [our achieving] this level of technology will be the ability to analyze and repair the human body as completely and effectively as we can repair any conventional machine today."

In *Engines of Creation*, Drexler drew inspiration from the cell's eye view to recognize that nanotechnology could bring a fundamental breakthrough in medicine. Noting that 20th century physicians relied chiefly on surgery and drugs to treat illness, Drexler explained:

"Surgeons have advanced from stitching wounds and amputating limbs to repairing hearts and reattaching limbs. Using microscopes and fine tools, they join delicate blood vessels and nerves. Yet even the best microsurgeon cannot cut and stitch finer tissue structures. Modern scalpels and sutures are simply too coarse for repairing capillaries, cells, and molecules. Consider 'delicate' surgery from a cell's perspective. A huge blade sweeps down, chopping blindly past and through the molecular machinery of a crowd of cells, slaughtering thousands. Later, a great obelisk plunges through the divided crowd, dragging a cable as wide as a freight train behind it to rope the crowd together again. From a cell's perspective, even the most delicate surgery, performed with exquisite knives and great skill, is still a butcher job. Only the ability of cells to abandon their dead, regroup, and multiply makes healing possible."

"Drug therapy, unlike surgery, deals with the finest structures in cells. Drug molecules are simple molecular devices. Many affect specific molecules in cells. Morphine molecules, for example, bind to certain receptor molecules in brain cells, affecting the neural impulses that signal pain. Insulin, beta blockers, and other drugs fit other receptors. But drug molecules work without direction. Once dumped into the body, they tumble and bump around in solution haphazardly until they bump a target molecule, fit, and stick, affecting its function. Drug molecules affect tissues at the molecular level, but they are too simple to sense, plan, and act. Molecular machines directed by nanocomputers will offer physicians another choice. They will combine sensors, programs, and molecular tools to form systems able to examine and repair the ultimate components of individual cells. They will bring surgical control to the molecular domain."

By the end of the 20th century, mainstream military (DoD), NIH, NSF, and other international groups had begun to seriously consider the potential future applications of molecular nanotechnology in medicine. For example, in 1997 a panel of U.S. Department of Defense health science experts known as Military Health Service Systems (MHSS) 2020 concluded in its final report:

"If a breakthrough to a [molecular] assembler occurs within ten to fifteen years, an entirely new field of nanomedicine will emerge by 2020. Initial applications will be focused outside the body in areas such as diagnostics and pharmaceutical manufacturing. The most powerful uses would eventually be within the body. Possible applications include programmable immune machines that travel through the bloodstream, supplementing the natural immune system; cell herding machines to stimulate rapid healing and tissue reconstruction; and cell repair machines to perform genetic surgery."

The present book takes as its starting point the assumption that the mass production of nanomachines at modest cost is technically feasible, and then explores the medical implications of this assumption. Proposed systems presented in this trilogy are intended not as final engineering blueprints but merely as points of departure for further analysis and refinement. All designs and projected capabilities are, for the most part, conservatively drawn with generous safety margins, leaving a considerable volume of design space yet to be explored by more intrepid future investigators.

Nanomedicine: History of the Idea

Conclusive proof of the existence of atoms was not obtained until the close of the 19th century. This may explain why the idea of nanomedicine is an exclusively 20th century phenomenon. The first hint of it may be found in a famous 1929 essay written by J.D. Bernal:

"The discoveries of the twentieth century, particularly the micro-mechanics of the Quantum Theory which touch on the nature of matter itself, are far more fundamental and must in time produce far more important results. The first step will be the development of new materials and new processes in which physics, chemistry and mechanics will be inextricably fused. The stage should soon be reached when materials can be produced which are not merely modifications of what nature has given us in the way of stones, metals, woods and fibers, but are made to specifications of a molecular architecture. Already we know all the varieties of atoms; we are beginning to know the forces that bind them together; soon we shall be doing this in a way to suit our own purposes. The result—not so very distant—will probably be the passing of the age of metals and all that it implies—mines, furnaces, and engines of massive construction. Instead we should have a world of fabric materials, light and elastic, strong only for the purposes for which they are being used, a world which will imitate the balanced perfection of a living body."

Development of the concept of nanomedicine has followed two principal paths which Richard Smalley has termed "wet nanotechnology" in the biological tradition, and "dry nanotechnology" in the mechanical tradition. Both approaches were presaged in speculative fiction. The following abbreviated history focuses on nanomedicine largely to the exclusion of broader issues in molecular engineering, manufacturing and nanoscopy, and includes a number of inspirational, speculative, or fictional early references from non-refereed sources.

Biological tradition

The general idea of biological engineering stretches back at least to the mid-19th century, but the first science fiction story involving actual genetic engineering was "Proteus Island" (1936), written by the chemical engineer Stanley Weinbaum. Artificial engineered organisms subsequently appeared in minor roles in several stories, a notable example being the familiars employed by the fake witches in Fritz Leiber's "Gather, Darkness!" (1943), and A.E. van Vogt used "gene transformation" to create the superman in "Slan" (1940). The first artificially evolved creatures appeared in Theodore Sturgeon's

"Microcosmic God" (1941), wherein a biochemist established conditions allowing accelerated artificial evolution, creating the Neoterics, a submillimeter-sized race of intelligent hypermetabolic creatures which could accomplish tasks very rapidly. The first artificially designed microcreatures appeared in James Blish's "Surface Tension" (1952). In this story, crash-landed dying human astronauts create a completely new form of humanity—tiny men and women reduced to protozoan size—who are seeded in the pools and puddles at the surface of the new planet, and who go on to master the biotechnology necessary to travel from one water puddle to another.

The scientific tradition of biological nanomachines for medical purposes began in 1964 when Robert Ettinger, an early cryonics pioneer, suggested that cellular-level or even molecular-level repair might be developed for life extension. Ettinger speculated that "...surgeon machines, working 24 hours a day for decades or even centuries, will tenderly restore the frozen brains, cell by cell, or even molecule by molecule in critical areas."

In 1965, the synthesis of artificial life was publicly proposed as a national goal by the president of the American Chemical Society, Professor Charles Price, who pointed out that many new types of life might be made, not "mere imitations" of biology as we know it.

In 1967, Isaac Asimov suggested the future possibility of "factories...where the working machinery consists of submicroscopic nucleic acids" and that a "repertoire of hundreds or thousands of complex enzymes" could be used to "bring about chemical reactions more conveniently than any methods now used" and also for "helping to construct life."

In 1968, G.R. Taylor cited the possibilities for genetic engineering and genetic surgery: "The microsurgery of DNA may possibly be achieved by physical methods: fine beams of radiation (probably laser light or pulsed X-rays) may be used to slice through the DNA molecule at desired points." He also cited predictions that bacteria would soon be programmed.

In 1969, J. White suggested that a modified virus could be used as a cell repair machine: "It has been proposed that appropriate genetic information be introduced by means of artificially constructed virus particles into a congenitally defective cell for remedy; similar means may be used for the more general case of repair. The repair program must use means such as protein synthesis and metabolic pathways to diagnose and repair any damage... [Information] can be preserved by specifying that the repair program incorporate appropriate RNA tapes into itself..."

In 1970, Jeon et al carried out "the reassembly of *Amoeba proteus* from its major components: namely nucleus, cytoplasm, and cell membrane," taken from three different cells.

In 1972, Ettinger proposed using genetic engineering to make microscopic biorobots: "Genetic engineering's most sensational impact will concern the modification of humans, but it will have other uses as well. Some of the "robots" that will serve us will need to be nanominiaturized." Existing organisms could be modified to make biologically-based programmable biorobots for medical applications: "If we can design sufficiently complex behaviour patterns into microscopically small organisms, there are obvious and endless possibilities, some of the most important in the medical area. Perhaps we can carry guardian and scavenger organisms in the blood, superior to the leukocytes and other agents of our human heritage, that will efficiently hunt down and clean out a wide variety of hostile or damaging invaders." Computerized cellular repair machines "must use means such as protein synthesis and metabolic pathways to diagnose and repair any damage...[Information] can be preserved by specifying that the repair program incorporate appropriate RNA tapes into itself...." Also in 1972, Danielli described various possibilities for generating new life forms via "life-synthesis" and genetic engineering, noting that "macromolecular engineering" might enable the development of very powerful and compact macromolecular computer systems.

In 1974, Halacy noted "some rather inglorious ways" to use "the miracle of artificial life," including potential capabilities for growing diverse items ranging from computers to airplanes. Morowitz suggested cooling cells to cryogenic temperatures in order to analyze and determine their structure. Artificial cells could also be assembled at such temperatures and then be set in motion by thawing. Morowitz further reported that microsurgery experiments on amoebas "have been most dramatic. Cell fractions from four different animals can be injected into the eviscerated ghost of a fifth amoeba, and a living functioning organism results."

In 1975, Richard Laing described the theoretical possibility of molecular machine self-replicators using molecular (data) tapes based on the idea of universal Turing machines, examining several ways that such "artificial organisms" might replicate themselves as "a vehicle for the exploration of broad biological possibilities."

In 1976, Donaldson presented the first detailed (and quite ambitious) list of biotechnological techniques that appeared necessary to achieve cell repair and might prove feasible, writing at a time that predated many current capabilities such as automated protein/DNA sequencing and synthesis, and most knowledge of restrictions on cellular developmental pathways and genetic programs and networks. For repair at the level of the cell, Donaldson's techniques would have included:

1. The ability to design enzymes to produce specific repair functions such as renaturing denatured proteins, joining broken lipoprotein complexes, annealing broken strands of DNA or RNA, reading proteins of existing or special types onto RNA and replicating them, and giving a cell the ability to metabolize new substrates, use novel cofactors, or construct essential amino acids;
2. Specially constructed bacteria or macrophages able to replicate themselves, spread throughout a specific target tissue, and carry out specific repairs according to the programs designed into their DNA/RNA; these could be designed to operate at unnatural temperatures or to utilize metabolic pathways not presently found in nature;
3. The abilities to re-introduce lost DNA or lost organelles such as mitochondria into a cell, to introduce entirely new forms of organelles perhaps to perform specific repair functions, and to introduce new metabolic capacities into a cell;
4. The ability to modify at will the developmental program of a cell, as for instance to induce postmitotic cells such as neurons to divide, according to a specific program, forming daughter cells with specific properties; and
5. Several different types of repair bacteria able to work together in an integrated fashion, and linked together by chemical means (e.g. hormones), so as to apply optimal repairs to every body cell in order to (A) diagnose the precise nature of the damage, call other repair bacteria to its location, and report to the attending doctor that new types of repair bacteria other than those already introduced are needed, and (B) identify structures which must be preserved (e.g. memory) and reconstruct them if necessary.

For repairs at the level of the whole organism, Donaldson offered the following rather aggressive biological nanotechnology techniques:

1. Understanding the physiology of aging combined with the ability to reverse it;

2. Control over growth and development, including the ability to program types of growth and development which do not naturally occur, such as growth of new eyes or other organs which have been lost or damaged, growth of an entire and well-formed body from a head alone, and regrowth of injured or lost brain tissue; and
3. Nonpermanent "substitute organs" which will take over from others which have been lost, including (A) the ability to keep a given tissue alive and healthy in vitro for an indefinite time and similar abilities for a body part, and (B) temporary replacements for any body organ which may have been lost. These would be used to support the body while new organs were growing, e.g. as "metabolic crutches." This capacity specifically includes the ability to make temporary replacements for diffuse systems such as the vascular or nervous systems. For instance, a specially created "plant" would grow an entire vascular system into the patient, down to replacement for the capillaries and venules from a single seed, always introducing its fibrils between cells and destroying none or very little of the original structure.

The last of these is a description of a (clearly speculative) "repair net," a concept which Donaldson may have been the first to propose in 1976. A whole-body repair net was later termed a "chrysalis", and twelve years later, in 1988, Donaldson provided an artist's conception and an additional description of these proposed biotechnological instrumentalities:

"Severe crushing or mangling injuries require us to provide a new vascular system. The repair device might resemble a fungus, growing mycelia into the injured tissue. A repair net would grow into a [crushed] limb, guided by recognition of the injured cells and a plan for how the limb should look after repair....[In the most severe injuries,] a chrysalis first envelops the patient, then enters in between all his cells. It disassembles the patient, surrounding each cell with its own repair machinery and vascular system. The geometry already preserves information about locations of the patient's cells. If necessary, morphogen chemical gradients could also retain this information. A patient would [locally] swell up to 10 times original diameter. After repair, the chrysalis withdraws the same way it entered."

In 1977, Darwin further developed the theme of tissue repair and cellular repair biomachines, independently proposing a modified white blood cell to perform repair functions.

In 1981, Asimov suggested that we "consider the bacteria. These are tiny living things made up of single cells far smaller than the cells in plants and animals....[We] can, by properly designing these tiniest slaves of ours, use them to reshape the world itself and build it close to our hearts' desire." More importantly, Donaldson elaborated on his earlier discussion of how cryonically suspended human beings might be repaired. He extended the earlier concepts of Ettinger and White, concluding that "with such hybrid technology as micro-miniature biological-mechanical machines the size of viruses" and related technology, "it seems unlikely that (to repair a single cell) there would be any difficulty at all in principle to carrying out any imaginable repair." He estimated that about 10 programmable cell-repair biomachines could be introduced into a cell that was under repair without causing too much mechanical disruption. In 1988, Donaldson described artificial macrophages that "can carry control machinery to recognize target cells, responding only to them, or responding differently depending upon cell type or cell conditions. They can still work even if the target cell isn't functioning (unlike viruses), rebuild target cell machinery other than the genes, and transfer many more genes up to an entire copy of the patient's genome." They can also "communicate with one another [and] release diffusible chemicals to guide one another's behaviour."

By the 1990s, bioengineered viruses of various types and certain other vectors were routinely being used in experimental genetic therapies as a means to target and penetrate certain cell populations, with the objective of inserting therapeutic DNA sequences into the nucleus of human target cells *in vivo*. Retrovirally-altered lymphocytes (T cells) began to be injected into humans for therapeutic purposes. Another example of an engineered cell in therapeutics was the use of genetically modified cerebral endothelial cell vectors to attack glioblastoma, which was being pursued by Neurotech (in Paris) in 1998. Engineered bacteria were being pursued by Vion Pharmaceuticals in collaboration with Yale University. In their "Tumour Amplified Protein Expression Therapy" program, antibiotic-sensitive *Salmonella typhimurium* (food poisoning) bacteria were attenuated by removing the genes that produce purines vital to bacterial growth. The tamed strain (cell line VNP20009) could not long survive in healthy tissue, but quickly multiplied ~1000-fold inside tumours, which are rich in purines. The next step would be to add genes to the bacterium to produce anticancer proteins that can shrink tumours, or to modify the bacteria to deliver various enzymes, genes, or prodrugs for tumour cell growth regulation. The engineered bacteria were available in multiple serotypes to avoid potential immune response in the host. Phase I human clinical trials were expected to begin in 1999 using clinical dosages produced in 50-liter fermenters, and other possible bacterial vector species were being examined.

Micro-biorobotics was still regarded by many in the biotechnology community as a highly speculative topic in 1998. Glen A. Evans described the possible construction of synthetic genomes and artificial organisms. His proposed strategy involves determining or designing the DNA sequence for the genome, synthesizing and assembling the genome, introducing the synthetic DNA into an enucleated pluripotent host cell, then introducing the host cell into an organism. Evans foresees the high throughput "read-out" of gene products discovered through genomic sequencing, rapid construction of designer genes and genomes, automated designer vector synthesis for gene replacement/insertional mutagenesis, and finally the construction of synthetic genomes and artificial organisms. Robert Bradbury has considered requirements and costs for genome synthesis and replacement. For example, Bradbury estimates genome synthesis costs by assuming a projected ~\$0.03/base for DNA synthesis (compared to ~\$0.80-\$1.00/base-micromole in 1999), with ~20,000 expressed genes per biorobotic cell with an average gene size of ~3000 bases, giving a ~60 megabase expressed genome per biorobotic cell (1 chromosome) and a raw synthesis cost of ~\$3 million per designer cell line (excluding design costs). There has also been discussion of "chemical reaction automata" as precursors for synthetic organisms, synthetic lifeforms and "nanobiology", "biorobotics", "cell rovers", microbial engineering, lymphocyte engineering, and artificial chromosomes, and cell engineering is rapidly gaining popularity.

Mechanical tradition

The late science fiction author Robert A. Heinlein nearly invented the concept of molecular nanotechnology in 1942 when he suggested a process for manipulating microscopic structures. Heinlein envisioned the extensive use of life-size teleoperator hands, called "waldoes," complete with sensory feedback for full, remote-controlled telepresence. His fictional hero, Waldo, used a collection of these mechanical teleoperated hands for building and operating a series of ever-smaller sets of such mechanical hands. The smallest mechanical hands, "hardly an eighth of an inch across," were equipped with micro-surgical instruments and stereo "scanners," and were used to "manipulate living nerve tissue, [to examine] its performance *in situ,*" and to perform neurosurgery. Eric Frank Russell's 1947 story "Hobbyist" described a fabrication process with "atom fed to atom like brick after brick to build a house." In 1955, Russell's serial "Call Him Dead," also published in *Astounding Science Fiction*, had a virus-based alien intelligence that spread through contact with blood or saliva; the story features a "microforger," a man who makes "surgical and manipulatory instruments so tiny they can be used to operate on a bacillus." Also in the mechanical tradition, Isaac Asimov's "Fantastic Voyage" in 1966 took its miniaturized human crew in a miniaturized submarine through the bloodstream of a human patient on a mission of repair.

Heinlein, Russell, and Asimov overlooked the full implications of their ideas, and most scientists were unsympathetic—for example, in 1952 Erwin Schrodinger wrote that we would never experiment with just one electron, atom, or molecule. But by the early 1960s, several scientists had reinvented similar approaches to micromanipulation and miniaturization, this time extending their reach into the nanotechnology domain. The first and most famous of these scientists was the Nobel physicist Richard P. Feynman. In his remarkably prescient 1959 talk "There's Plenty of Room at the Bottom," Feynman proposed employing machine tools to make smaller machine tools, these to be used in turn to make still smaller machine tools, and so on all the way down to the atomic level. Feynman prophetically concluded that this is "a development which I think cannot be avoided." Such nanomachine tools, nanorobots and nanodevices could ultimately be used to develop a wide range of submicron instrumentation and manufacturing tools, i.e., nanotechnology. Feynman's suggested applications for these tools included producing vast quantities of ultrasmall computers and various micro- and nano-robots.

Feynman was clearly aware of the potential medical applications of the new technology he was proposing. After discussing his ideas with a colleague, Feynman offered the first known proposal for a nanomedical procedure to cure heart disease: "A friend of mine suggests a very interesting possibility for relatively small machines. He says that, although it is a very wild idea, it would be interesting in surgery if you could swallow the surgeon. You put the mechanical surgeon inside the blood vessel and it goes into the heart and looks around. (Of course the information has to be fed out.) It finds out which valve is the faulty one and takes a little knife and slices it out. Other small machines might be permanently incorporated in the body to assist some inadequately functioning organ." Later in his historic lecture, Feynman urged us to consider the possibility, in connection with biological cells, "that we can manufacture an object that maneuvers at that level!"

In 1961, K.R. Shoulders rejected the use of biological building blocks, even though biological "processes do work, and they can do so in a garbage can without supervision." He saw them as too limited environmentally and too difficult to control with available technology. Instead, he sought to directly produce much simpler, more powerful and rugged nanostructure arrays, operating at video frequency rates, which in turn could ultimately aid in their own replication. In 1965, Shoulders reported the actual operation of micromanipulators able to position tiny items with 10 nm accuracy while under direct observation by field ion microscopy.

In 1970, Volkenstein noted that "the creation of a nonmacromolecular system which would act as a model for living organisms is definitely possible" but could not arise by itself, and that the macromolecularity of present organisms is not essential, but due to their evolutionary origins. Taking some poetic license, he added: "Consequently, the cybernetic nonmacromolecular machine, which simulates life, could have been and can be created on earth only by man. Then it could perfect itself without limits." T. Nemes discussed artificial self-replicating machines and described how to construct "an automatic lathe able to reproduce itself," a concept apparently developed before von Neumann's work on machine replication.

In 1981, Drexler suggested the construction of mechanically deterministic nanodevices using biological parts; these devices could inspect cells at the molecular level and also repair cellular tissues that had been damaged during cryonic suspension. In 1982, Drexler described cell repair machines even more clearly in the mechanical tradition, in a popular publication.

By 1983, Drexler began privately circulating a draft technical paper entitled "Cell Repair Machines" which investigated for the first time, in some detail, whether an advanced mechanical-based nanotechnology would "permit construction of systems of molecular-scale sensors, computers, and manipulators able to enter and repair cells; the nature of the computational algorithms and magnitude of the computational resources needed to guide repairs; and the physical capabilities and constraints important to the repair process [in order to] sketch the conceptual design of a cell repair system based on a mature molecular technology."

Peterson notes that "medical applications were explored by Drexler during the early 1980s but the medical community was not ready for the concept." A technical paper by Drexler, invited by an editor at the *Journal of the American Medical Association*, was dismissed by a referee as "science fiction."

In 1985, G. Feinberg proposed using short-wavelength coherent laser energy to power and communicate with "nanosensors that could be implanted into the human body. They could...[monitor] various physiological functions from subcellular molecules up through tissues and organs...essential in determining some of the mechanisms involved in growth and aging."

In a 1985 book entitled *Robotics*, edited by Marvin Minsky (a well-known computer scientist and artificial intelligence pioneer), Minsky briefly described how fully-automatic cellular repair machines might work, following Drexler's vision:

"Suppose that we could design a repair machine so small that it could repair an artery from the inside! The first such machines might be the size of fleas. (There is room for a great deal of machinery in something the size of a flea—as the body of the flea itself testifies.) These micromachines could crawl into all but the smallest blood vessels, clear out debris, and reline the walls with suitable materials yet to be invented. Later generations of micromachines could be even smaller, perhaps no larger than the body cells they repair. These minuscule machines would be mass-produced by the billions, either by larger machines or by techniques of making them reproduce themselves. Perhaps these biological janitors would even be implanted in our bodies to remain there as permanent maintenance workers, just like many biological cells that already serve such purposes."

"Today the idea of such a technology may seem fantastic, yet many of the circuits in our computers are already smaller than many of our bodies' cells. Let's try, for a moment, to look ahead to: mass production of highly intelligent machines; gigantic advances in miniaturization; a technology so advanced that these machines reproduce themselves without our help. Fantastic? Not at all. Even the simplest algae and bacteria can do that. True, they're not intelligent, but each of them contains enough computerlike machinery and memory to do those things. So, to build bacteria-size computers should be perfectly feasible, once we have the necessary microtechnology. Some day we'll have the means to build artificial, cell-like machines with all those capabilities."

R.A. Freitas Jr. also contributed a chapter to Minsky's 1985 book, offering the following suggestion for remote-controlled incisionless nanosurgery: "One possibility is the concept of remote-controlled medical mites made feasible by modern micromachinery technology. Some medical mites would be like microminiature submarines, released inside the human body for internal sensing. Other mites could float, crawl, or swim through major arteries in the human body and perform on-site repairs from within, controlled by radio link under direction of a skilled telemicrosurgeon."

In 1986, Drexler published *Engines of Creation*, a popular text with two chapters devoted to discussions of cellular repair machines. Drexler further expounded upon the topic of cellular repair machines in articles published in 1985, 1986, 1987, and 1989, and his concepts were described by Brian Wowk in 1988 and in a Time-Life book in 1989.

In 1988, A.K. Dewdney reported an early nanomedical concept of an artery-cleaning nanorobot that he attributed to Drexler, accompanied by an artist's conception with the caption "a nanomachine swimming through a capillary attacks a fat deposit."

Nanomedicine in the 1990s

Despite the tremendous importance of molecular nanotechnology, a published 1993 literature review of robotics in health care included not a single reference to nanotechnology or to nanomedicine. At this writing in 1998, the list of original post-1989 technical and popular nonfiction works that deal with nanomedical topics in the mechanical tradition is short enough to permit virtually an exhaustive listing, including Beardsley, Bova, Coombs and Robinson, Crawford, Drexler, DuCharme, Emanuelson, Fahy, Fiedler and Reynolds, Freitas, Kaehler, Klatz and Kahn, Kurzweil, Lampton, Merkle, Merrill, Minsky, Morc, Ostman, Reifman, and Wowk and Darwin. The first nanomedical device design technical paper was published in 1998 by Freitas in the biotechnology journal *Artificial Cells*, and the present trilogy (*Nanomedicine*) is the first book-length technical treatment of the medical implications of molecular nanotechnology.

Biotechnology and Molecular Nanotechnology

Some pragmatic readers may be wondering what is so special about molecular nanotechnology, that existing or anticipated biotechnology could not accomplish just as well? After all, biotechnology is already an established medical capability. It has real applications and real products already on the market. Reflecting upon the future possibility of sophisticated mechanical medical nanorobots equipped with powerful nanocomputers, in 1989, one well-known cryobiologist mused that "what is not clear is just what need we would have for such devices." The simplest answer, is that each of the three contemporary branches of "nanotechnology" offers something of unique value to the practice of medicine.

Nanoscale materials technology has already found widespread use in medicine, including biocompatible materials and analytical techniques, surgical and dental practice, nerve cell research using intracellular electrodes, biostructures research and biomolecular research using near-field optical microscopy, scanning-probe microscopy and optical tweezers, and vaccine design, and also many 20th century bulk chemical and biochemical manufacturing techniques along with much of classical pharmacology.

As for "biotechnology," the original meaning of this word contemplated "the application of biological systems and organisms to technical and industrial processes". In recent times, the field has expanded to include genetic engineering and now takes as its ultimate goal no less than the engineering of all biological systems, even completely artificial organic living systems, using biological instrumentalities.

The third branch, molecular nanotechnology, takes as its purview the engineering of all complex mechanical systems constructed from the molecular level—potentially offering new tools for medical practice, the principal subject of this book. Observes G.M. Fahy, "the difference between nanotechnologists and biotechnologists is that the former do not restrict themselves to the biological limitations of the latter, and they are much more ambitious about the kinds of accomplishments that they want to achieve."

Doctors can utilize solutions to medical problems from all three approaches. As noted earlier, 80-90% of medical complaints resolve themselves via natural homeostatic processes, or without the necessary involvement of active biotechnological or molecular-nanotechnological agents. But by employing biotechnology, the range and efficacy of treatment options greatly increases. With molecular nanotechnology, the range, efficacy, comfort and speed of possible medical treatments again expands enormously. Molecular nanotechnology is essential when the damage to the human body is extremely subtle, highly selective, or time-critical (as in head traumas, burns, or fast-spreading diseases), or when the damage is very massive, overwhelming the body's natural defenses and repair mechanisms.

At every difficulty level, most classes of medical problems may be resolved with varying efficacy within the homeostatic/nanomaterials, biotechnological, or molecular-nanotechnological approaches. As the chosen technology becomes more precise, active, and controllable, the range of options broadens and the quality of the options improves. Thus the question is not whether molecular nanotechnology is required to accomplish a given medical objective. In many cases, it is not—though of course there are some things that only biotechnology and molecular nanotechnology can do, and some other things that only molecular nanotechnology can do. Rather, the important question is which approach offers a superior solution to a given medical problem, using any reasonable metric of treatment efficacy. For virtually every class of medical challenge, a mature molecular nanotechnology offers a wider and more effective range of treatment options than any other approach.

It is quite possible to imagine an advanced biotechnology that uses an engineered white cell, fibroblast, or macrophage chassis, energized by native oxygen and glucose and modified mitochondrial powerplants, driven by pseudopodia, cilia or flagella, communicating and navigating via biochemical signals, and even incorporating onboard digital biocomputers to make microscopic biorobots. Principal arguments favouring the biotechnology approach for medical purposes are: (1) that we are already somewhat familiar with such systems, after half a century of intensive molecular biology research; (2) that we have already "built" precursor systems, such as a whole living amoeba constructed from five distinct parts, bioengineered viruses and bacteria as DNA insertion devices, and natural replication stimulated in genetically engineered starter microbes; (3) that biocompatibility will not be a major issue since fibroblasts (which express no HLA Class II antigens, hence stimulate no rejection response) could be used as the starter material; (4) that both engineered viruses and bacteria are already in wide commercial and research use; and (5) the greater complexity of self-repair in mechanical systems, should it be needed. Many believe that the development pathway to early biorobots may be considerably shorter than for the mechanical nanorobots of the molecular nanotechnology approach, for which in 1998 not a single working prototype yet existed, even in research laboratories.

It is also possible to imagine a molecular nanotechnology that uses mechanical nanorobotic systems. Such systems will have many constitutional differences from biological-based systems. For instance, mechanical systems will transport parts, materials, energy and instructions via fixed channels, whereas most (but not all) biological systems operate by diffusion. Mechanical systems will have structures constrained by specific geometries, whereas biological systems have structures defined by patterns of containment and interconnection—the shape of a membrane compartment in a cell matters less than its continuity and the contents of the volume it defines. Mechanical systems will be deterministically manufactured by operations analogous to manual construction, whereas engineered ribosomes will self-assemble via diffusion and stochastic matching of complementary parts; in other words, biology uses recipes, while mechanical systems use blueprints. Cells grow, with their parts adapting to one another; mechanical nanorobots may be constructed from parts of fixed structure. Biology uses self-repair; mechanical systems generally do not largely because self-repair (by component exchange) will not be needed in molecular mechanical systems, whose designs may be made more simple by relying upon high component redundancy. These and other differences imply a number of important advantages that mechanical-based medical systems will enjoy over biological-based medical systems, which, taken together, strongly suggest that predominantly mechanical nanosystems may be the approach of choice for a mature medical nanotechnology. The advantages of molecular nanotechnology (e.g. the mechanical tradition) are many.

Speed of medical treatment

Doctors may be surprised by the incredible quickness of nanorobotic action when compared to the speeds available from fibroblasts or leukocytes. Normal homeostatic processes such as dermal wound repair via natural fibroblasts may require weeks to run to completion. Typical fibroblast movements occur at 0.1-1 microns/sec, but mechanical nanomanipulators can operate at 1-10 cm/sec speeds or faster, a speed advantage of 4-5 orders of magnitude. Even the strongest biological fibers (e.g. intermediate filaments) have a failure strength 3 orders of magnitude below the strongest mechanical fibers (e.g. fullerene nanotubes). Biological cilia beat at ~30 Hz while mechanical nanocilia may cycle up to ~20 MHz, though practical power restrictions and other considerations may limit them to the ~10 KHz range for most of the time. Flexible mechanical surfaces can complete a morphing motion in ~0.1 millisec, compared to the ~100 millisec snapback time for pinched red cell membrane, again a thousand-fold advantage in speed. Thus we expect that mechanical therapeutic systems can reach their targets up to ~10,000 times faster, all else equal, and treatments which require ~10^5 sec for a biological system may need only ~10^2 sec for a mechanical system; tachyiatria improves both patient and physician comfort. In either biological or mechanical systems, large numbers of devices of comparable physical size (e.g. ~microns) may be employed to do the work, so numbers alone cannot offset the mechanical speed advantage.

Power density and transduction

Biological cells typically employ power densities of 10^3-10^4 W/m^3, with maximum densities of ~10^6 W/m^3 in honeybee flight muscle cells and bacterial flagellar motors. By contrast, nanomechanical power systems can produce power densities of 10^9-10^{12} W/m^3, an advantage of 10^3-10^8 for mechanical over biological systems. By 1998, conducting polymer-based actuators generated 20-100 times the force for a given cross-sectional area as mammalian skeletal muscle. Additionally, amoebic locomotion in motile cells requires diffusion-limited cytoskeletal disassembly and reassembly to achieve movement; mechanical motility systems may employ simple cable-pulling, winches, or ratchets, which are faster and more direct. Some biological energy transducers are reversible, but muscle contraction is irreversible—not only cannot muscle actively re-expand, but stretching it doesn't make it produce much useful chemical energy. In contrast, electric motors can be run backwards to generate electricity, forcing pistons makes them pump, and loudspeakers can be used as microphones.

Superior building materials

Typical biological materials have tensile failure strengths in the 10^6-10^7 N/m^2 range, with the strongest biological materials such as wet compact bone having a failure strength of ~10^8 N/m^2, all of which compare poorly to ~10^9 N/m^2 for good steel, ~10^{10} N/m^2 for sapphire, and ~10^{11} N/m^2 for diamond and carbon fullerenes, again showing a 10^3-10^5 advantage for mechanical systems that use nonbiological materials. Nonbiological materials can be much stiffer, permitting the application of higher forces with greater precision of movement, and they also tend to remain stable over a larger range of temperature, pressure, salinity and pH. Proteins are heat sensitive in part because much of the functionality of their structure is due to noncovalent bonds involved in folding, which are broken more easily at higher temperatures; in diamond, sapphire, and many other rigid materials, structural shape is covalently fixed, hence is far more temperature-stable. Most proteins tend to become dysfunctional at cryogenic temperatures, unlike diamond-based mechanical structures. Biomaterials are not ruled out for all nanomechanical systems, but represent only a small subset of the materials that can be used in nanorobots. Mechanical systems can employ a wider variety of atoms and molecular structures in their design and construction, with novel functional forms that might be difficult to implement in a biological system such as steam engines or nuclear power. As another example, an application requiring the most effective bulk thermal conduction possible should use diamond, the best conductor available, not some biomaterial with inferior thermal performance.

Nondegradation of treatment agents

Diagnostic and therapeutic agents constructed of biomaterials generally are biodegradable in vivo, although there is a major branch of pharmacology devoted to designing drugs that are moderately non-biodegradable—anti-sense DNA analogues with unusual backbone linkages and peptide nucleic acids (PNAs) are difficult to break down. However, suitably designed nanorobotic agents constructed of nonbiological materials are not biodegradable. An engineered fibroblast may not stimulate an immune response when transplanted into a foreign host, but its biomolecules are subject to chemical attack *in vivo* by free radicals, acids, and enzymes. Even "mirror" biomolecules or "Doppelganger proteins" comprised exclusively of unnatural D-amino acids have a lifetime of only ~5 days inside the human body. Nonbiological materials such as diamond and sapphire are highly resistant to chemical breakdown or leukocytic degradation *in vivo*, and pathogenic biological entities cannot easily evolve useful attack strategies against these materials.

Control of nanomedical treatment

Present-day biotechnological entities are not programmable and cannot be switched on and off conditionally during task execution. A digital biocomputer, while possible in theory, represents a considerable conceptual departure from the usual biological paradigm. Even assuming that a digital biocomputer could be installed in a fibroblast, and that appropriate effector mechanisms could be attached, such a system would necessarily have slower clock cycles, less capacious memory per unit volume, and longer data access times, implying less diversity of action, poorer control, and less complex executable programs than would be available in nanoscale electromechanical computer systems. The mechanical approach emphasizes precise control of action, including control of physical placement, timing, strength, structure, and interactions with other (especially biological) entities. The biological approach emphasizes the use of poorly controlled natural structures, needlessly sacrificing huge blocks of the available functionality and design space.

Nanodevice versatility

Mechanical systems can readily incorporate biological elements if necessary, but artificial biological systems can incorporate nonbiological materials such as carbon nanotubes or diamond/sapphire structural elements only with difficulty, in part because biology has a more limited repertoire of "effector" mechanisms. Artificial biological systems cannot easily incorporate nonbiological materials where desired because natural biological assembly methods make no provisions for these materials either in the coded instructions in DNA or in the attachment chemistries. Rebuilding or reconstructing the human body with nonbiological components (e.g. fullerene encabled bone for bone damaged in an accident), or augmentation of human body function with unnatural abilities or features (e.g. autogenous paracrine control), will be very difficult or impossible to achieve using purely biotechnological means.

Avoiding overspecialization

M. Krummenacker notes that one of the most glaring shortcomings of bacteria and other naturally occurring molecular machinery—when viewed as systems subject to further engineering—is the rather limited range of molecular substrates they can utilize. For example, bacterial enzymes are highly specialized devices, with very narrow substrate specificities. Thousands of different enzymes are needed in each organism, and the substance classes capable of digestion are limited to some sugars, various amino acids including proteins that can be degraded by excreted proteases, lipids, and a few other smallish oxygen-functionalized carbon molecules such as glycerol and ethanol. Some bacteria can metabolize CO_2 and a handful of aromatic compounds, but there is a vast range of organic chemicals that most bacteria cannot degrade or manufacture. A much smaller set of substantially more general molecular tools can probably be designed using the mechanical approach; mechanosynthesis can fabricate and assemble, or disassemble, a far wider range of molecular structures than are available to the cellular machinery of life.

Faster and more precise diagnosis

The analytic function of medical diagnosis requires rapid communication between the injected devices and the attending physician. If limited to chemical messaging, biotechnology devices will require minutes or hours to complete each diagnostic loop. Nanomachines, with their more diverse set of input-output mechanisms, can outmessage the results of *in vivo* reconnaissance or testing literally in seconds. Such nanomachines can also run more tests of greater variety in less time. Mechanical nanoinstruments, including molecule-by-molecule disassemblers, will make comprehensive cell mapping and cell interaction analysis possible. Bacterial resistance can be assayed at the molecular level, allowing new treatment agents to more easily be composed, manufactured and immediately deployed.

More sensitive response threshold for high-speed action

Unlike natural systems, an entire population of nanobiotic devices can be triggered globally by just a single local detection of the target antigen or pathogen. The natural immune system takes $>10^5$ sec to become fully engaged after exposure to a systemic pathogen or other antigen-presenting intruder. A biotechnologically enhanced immune system that can employ the fastest natural unit replication time ($\sim 10^3$ sec for some bacteria) will require $\sim 10^4$ sec for full deployment post-exposure. By contrast, a nanobiotic immune system can probably be fully engaged (though not finished) in at most two blood circulation times, or $\sim 10^2$ sec.

More reliable operation

Engineered macrophages would probably individually operate less reliably than mechanical nanorobots. Many pathogens, such as *Listeria monocytogenes* and *Trypanosoma cruzi*, are known to be able to escape from phagocytic vacuoles into the cytoplasm; while biotech drugs or cell manufactured proteins could be developed to prevent this (e.g. cold therapy drugs are entry-point blockers), nanorobotic trapping mechanisms can be more secure. Proteins assembled by natural ribosomes typically incorporate one error per $\sim 10^4$ amino acids placed; current gene and protein synthesizing machines utilizing biotechnological processes have similar error rates. A molecular nanotechnology approach will improve error rates by at least a millionfold. Mechanical systems can also incorporate sensors to determine if and when a particular task needs to be done, or when a task has been completed. Finally, it is unlikely that natural organisms will be able to infiltrate mechanical nanorobots or to co-opt their functions. By contrast, a

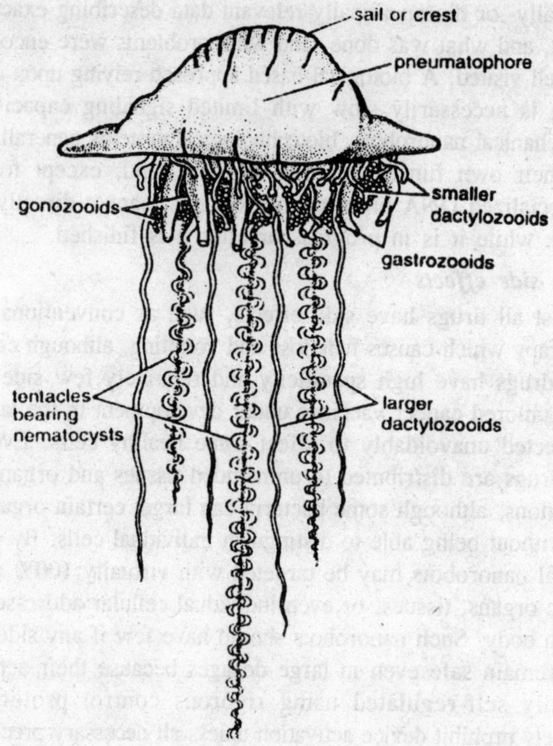

Fig. 1.5. Portuguese man-of-war.

biological-based robot could be diverted or defeated by microbes that can piggyback on its metabolism, interfere with its normal workings, or even incorporate the device wholesale into their own structures, causing the engineered biomachine to perform some new or different function than was originally intended. There are many examples of such co-option among natural biological systems, including the protozoan mixotrichs found in the termite gut that have assimilated bacteria into their bodies for use as motive engines, and the nudibranch mollusks (marine snails without shells) that steal nematocysts (stinging cells) away from coelenterates such as jellyfish (i.e. a Portuguese man-of-war) and incorporate the stingers as defensive armaments in their own skins, a process which S. Vogel has called "stealing loaded guns from the army."

Verification of progress and treatment

Using a variety of communication modalities, nanorobots can report back to the attending physician, with digital precision, a summary of diagnostically- or therapeutically-relevant data describing exactly what was found, and what was done, and what problems were encountered, in every cell visited. A biological-based approach relying upon chemical messaging is necessarily slow with limited signaling capacity. Also unlike mechanical nanorobots, biotechnological systems generally cannot monitor their own functions while working, and, except for a few highly specialized DNA proofreading systems, cannot directly inspect their work while it is in progress or after it is finished.

Minimum side effects

Almost all drugs have side effects, such as conventional cancer chemotherapy which causes hair loss and vomiting, although computer-designed drugs have high specificity and relatively few side effects. Carefully tailored cancer vaccines under development in the late 1990s were expected unavoidably to affect some healthy cells. Even well-targeted drugs are distributed to unintended tissues and organs in low concentrations, although some bacteria can target certain organs fairly reliably without being able to distinguish individual cells. By contrast, mechanical nanorobots may be targeted with virtually 100% accuracy to specific organs, tissues, or even individual cellular addresses within the human body. Such nanorobots should have few if any side effects, and will remain safe even in large dosages because their actions can be digitally self-regulated using rigorous control protocols that affirmatively prohibit device activation unless all necessary preconditions have been, and continuously remain, satisfied. G.M. Fahy has noted that these possibilities could transform "drugs" into "programmable machines with a range of sensory, decision-making, and effector capabilities [that] might avoid side effects and allergic reactions... attaining almost complete specificity of action....Designed smart pharmaceuticals might activate themselves only when, where, and if needed." Additionally, nanorobots may be programmed to excuse themselves from the site of action, or even from the body, after a treatment is completed; by contrast, spent biorobotic elements containing ingested foreign materials may have more limited post-treatment mobility, thus lingering at the worksite causing inflammation when naturally degraded or removed.

Reduced replicator danger

Drexler points out that living systems are evolved systems, while nanomechanical replicators would be designed: "The former are shaped to serve the goal of their own survival and replication in a natural environment, whereas the latter will be shaped (whether well or poorly) to serve human goals, perhaps in an artificial environment." Genetic engineering involves not design of replicators from scratch, but tinkering with the molecular machinery of existing bioreplicators. Since bioreplicators were not designed, they are not necessarily structured in ways that lend themselves to complete understanding, and processes based on diffusion and matching allow complex nonlocal interactions that can be hard to trace. Bioreplicators can be crippled, but having evolved in nature, they resemble systems that can survive in nature. Typically, they are able to exchange genetic information with wild organisms, raising the possibility of the introduction of new, unconstrained replicators in the natural environment. Having evolved to evolve, they have a capacity for further evolution—to serve their own survival, not human goals. For example, even when stripped of key pieces of DNA to interfere with its replication powers, a live attenuated AIDS vaccine can slowly recover its virulence and can attack immune cells. R. Bradbury suggests that artificial biorobots could incorporate multiple fail-safe mechanisms including required external essential nutrients or suicide suppressors, self-destruct triggers, countdown timers like telomeres, and engineering to reject foreign DNA (the basis for restriction enzymes), but it remains logically easier for a system with inchoate capacity to evolve to resume doing so, than for a system which has never had this capability to spontaneously develop it.

In contrast, nanomechanical replicators (e.g. assemblers) will be designed from scratch and thus will differ fundamentally from biological systems. The parts and structures of designed mechanical systems will be known, and the relationships among their parts will also be designed and fixed. More important, nanoreplicators will be fundamentally alien to the biosphere, unrelated to anything that has evolved to survive in nature. Certainly the capacity to fail can appear by accident, and emergent capabilities cannot be completely ruled out, but engineering experience shows that the ability to perform complex organized activities (such as replication in a natural environment) does not normally appear spontaneously. Also, and purely as a geometrical consideration, adding a new part inside a densely organized geometric structure (as would be found in a nanomachine) typically requires changes in the relative positions of many other parts, and hence corresponding adjustments in design. Adding a part inside a densely organized topological structure (as found in living systems) typically leaves topologies unchanged—room can be made by stretching and shifting other parts, with no change in their essential design, hence permitting easier modification to biological structures by exogenous agencies.

Note that mechanical medical nanodevices need not be capable of replication. There is no requirement for replication *in vivo*; such replication would be needlessly dangerous, and adding this capability would reduce effectiveness in carrying out the primary medical task. Analogously, viral vectors employed in genetic therapies are modified to be "incapable" of replication.

Assured patentability

Microscopic biorobots, unavoidably derived from natural biological material, may someday be deemed unpatentable under a general prohibition on "genetic colonialism" or other emerging legal doctrines. In contrast, mechanical nanorobots, being fully-artificial and designed machines, should always be patentable provided they satisfy the customary legal criteria.

Naturophilia

As already noted, living things in general and the human body in particular are awesome examples of a powerful and intricately woven natural molecular technology to which human engineers, in 1998, still aspire. But embracing Nature is not the same as finding her perfect. Today, the word "natural" has acquired a strong connotation of rightness, even of sanctity. For most of human history, notes biologist Steven Vogel, "the natural and human worlds stood opposed. Nature was something to be tamed and utilized; we had the ordinary attitude of organisms toward other species. Nowadays the natural world intrudes far less but gets venerated far more. And why not? When one's meat is bought in a store, when locusts don't threaten one's corn crop, when central heating and plumbing are the norm, the aesthetics of nature hold greater appeal." And so we embrace the natural rectitude or moral superiority of nature's ways, a kind of pantheism which may be called ethical naturalism, biophilia, or naturophilia.

Many great minds have fallen prey to naturophilia. In the 4th century BC, Aristotle wrote that "if one way be better than another, that you may be sure is Nature's way." In the 15th century, we have from Leonardo da Vinci: "Human ingenuity may make various inventions, but it will never devise any inventions more beautiful, nor more simple, nor more to the purpose than Nature does; because in her inventions nothing is wanting and nothing is superfluous."

However, as Virginia Postrel notes in *The Future and Its Enemies*: "If nature is itself a dynamic process rather than a static end, then there is no single form of "the natural.' An evolving, open-ended nature may impose practical constraints, but it cannot dictate eternal standards. It cannot determine what is good. The distinction between the artificial and the natural must lie not in their source—human or not—but in their characteristics, in the way they relate to the world around them."

According to the dictionary, "artificial" usually means "made by man, rather than occurring in nature." More usefully, Herbert Simon defines the artificial as that which is designed, expresses goals, and possesses external purposes. The artificial is controlled and serves its creators' purposes, subject to the universal laws of physics. Kevin Kelly defines the natural as "out of control." Nature is evolved, not designed, and serves no goal or external purpose save its own survival. Nature, lacking intent, is amoral—it simply is. By building the artificial, observes Postrel, "we do not overthrow nature, but cooperate with it, using nature's own art to create new natural forms. Our artifice alters the path of nature, but it does not end it, for nature has no stopping point, no final shape. It is a process, not an end."

Some naturophilist writers have decried the increasing "medicalization of society" in which formerly natural functions have come to be regarded as medical conditions requiring intervention or treatment. However, history suggests that naturophilia is usually undermined by any new medical technology that offers clear, safe, and immediate benefits to patients. For example, prior to 1842, intense pain was viewed as the natural outcome of being cut with a scalpel during surgery. It had always been so—how could it ever be otherwise? The invention of anesthesia in 1842 suddenly altered this natural outcome and replaced it with a less painful artificial outcome, despite anguished cries from naturophiles within the medical community that eliminating pain might somehow diminish the human character.

Another example of a widely-accepted medicalization of normal function is childbirth, a quite natural activity that can nevertheless be very dangerous to the mother's health. Precise prehistoric death rates are unknown, although archeological evidence shows that Neandertal females tended to die before the age of 30 due to hazards of childbirth. In the worst 19th century maternity hospitals the natural death rate from childbirth was 9-10%, falling to a very artificial 0.4% rate in England by 1930 and to less than 0.01% in the U.S. during the 1990s. As a result, it now seems "natural" for a woman always to survive childbirth, even though the reverse may have been true for most of human history. Warns historian Roy Porter: "We should certainly not hanker after some mythic golden age when women gave birth naturally, painlessly, and safely; the most appalling Western maternal death rate today is among the Faith Assembly religious sect in Indiana, who reject orthodox medicine and practice home births; their perinatal mortality is 92 times greater than in Indiana as a whole."

A disease seems "natural" to those who suffer from it when no treatment exists. But once a treatment is discovered and is widely employed, the disease becomes rare and its absence now becomes "natural." To those in the past, writes K.E. Drexler, "the idea of cutting people open with knives painlessly would have seemed miraculous, but surgical anesthesia is now routine. Likewise with bacterial infections and antibiotics, with the eradication of smallpox, and the vaccine for polio: each tamed a deadly terror, and each is now half-forgotten history. What amazes one generation seems obvious and even boring to the next. The first baby born after each breakthrough grows up wondering what all the excitement was about." In the next century, says Charles Sheffield, "our descendants will look on angiograms, upper and lower GIs, and biopsies the way we regard the prospect of surgery without anesthetics."

Future generations who take for granted an all-pervasive nanomedicine in their lives may look aghast upon the 20th century, wondering among other things how we managed to retain productive focus given the constant annoyance of our numerous undiagnosed minor disease states. Most of these diseases are not yet recognized as such, and many are still regarded as "natural" and not worthy of treatment. In a few decades, this may change. Some examples:

Addictions

In 1998, many people laugh off seemingly harmless addictions to chocolate (*chocoholics*), fats or sugar (*sweet tooth*), food (*gluttony*), nicotine (*smokers*), caffeine (*coffee* and *cola drinkers*), work (*workaholics*), exercise (*runner's high*), telling falsehoods (*pathological liars*), gambling (*wagerphilia*), stealing (*kleptomania*), medical treatments (*hypochondria*), marriage (*polygamy*), power-seeking (*domination*), skydiving or bungee-jumping (*thrillseeking*), superstition (*astrology*), shopping (*spendoholics*), driving cars that kill 40,000 Americans per year (*mobilophilia*), unusual sexual preferences (bestiality), sexual activity (*nymphomania*, *satyriasis*), or pregnancy (*gravidophilia*). Without making any value judgements, it is highly likely that most or all of these addictions have genetic or physiological components which, once properly modified, can greatly reduce or eliminate the addiction if so desired. Many on the list are already suspected to have genetic components, much like *schizophrenia*, drug abuse, *bulimia*, and *alcoholism* (*dipsomania*).

Allergies and intolerances

A food allergy is an allergic reaction to a particular food, although true food allergies are much rarer than is generally believed. In the cases of milk, eggs, shellfish, nuts, wheat, soybeans, and chocolate, sufferers may lack an enzyme necessary for digesting the substance. In other cases, dust particles, plant pollens, pet danders, drugs, or foods may be allergens for natural IgE-mediated immunosensitivity. Intolerance, a much more common condition, is any undesirable effect of eating a particular food, including gastrointestinal distress, gas, nausea, diarrhea, or other problems. *Urticaria* (hives), *angioedema* and even mild *anaphylaxis* are common reactions to various drugs, insect stings or bites, allergy shots, or certain foods, particularly eggs, shellfish, nuts and fruits. Physical allergies to ordinary stimuli such as cold, sunlight, heat, pollen, pet dander, or minor injury can produce itching, skin blotches, pimples, and hives.

Minor physical annoyances

In a world where most major medical maladies are readily treated, numerous minor medical conditions which today escape our notice will rise up from obscurity and present themselves annoyingly to our conscious minds, demanding attention. These conditions may be of several kinds. First is cosmetics, including small moles, freckles and blemishes on the skin; broken fingernails or unevenly-growing cuticles; minor skin reddenings or pimples; old childhood scars, wrinkled skin, birthmarks or stretch marks; unwanted hair growth in unusual places, or differential hair colour or texture growing in patches; fingerprint patterns that are aesthetically unappealing; and mismatched leg lengths, hands with different left/right ring sizes, an asymmetrical face, or lopsided breasts. Second is minor aches and pains, which may include headaches; eyebrow hairs trapped in the eyeball conjunctiva; bent-hair pain (*folliculalgia*); dyspepsia; creaking limb joints and stomach growling; ingrown nails and hairs; earwax plugs and temporary tinnitis; chapped lips, canker sores and heat rashes; stuffy nose or gritty eyes upon rising in the morning; dermal chafing marks from elastic bands in clothing; minor flatulence; PMS (premenstrual syndrome); a leg or arm "falling asleep" in certain postures; nervous tics, itches, and twitches; uncracked knuckles, stiff neck, or backache; blocked middle ear following descent from high altitude; restless leg syndrome (*akathisia*); rotationally-induced dizziness, as on an amusement park ride; or rock-and-roll neck, wherein active musical performers or listeners bob their heads violently, rupturing small blood vessels in the neck. Third is minor physical or functional flaws, such as poor stream during male urination, female papillary leakage, colourblindness, snoring, unpleasant body odors, nosebleeds, declining visual or aural acuity, handedness (currently ~90% dextromanual, ~10% sinistromanual, mild strabismus (eyeball misalignment), bad moods (neurotransmitter imbalances), or post-intoxication hangover.

Undiscovered infectious agents

Peptic ulcers once were thought to result from a stressful life, a purely natural response to a lifestyle choice. Then it was found that the major cause of ulcers is the presence of *Helicobacter pylori* bacteria in the stomach. Bacteria have been implicated in some cases of atherosclerosis and *Alzheimer's disease*, and nanobacteria have been proposed as possible nucleation sites for kidney stones. Other seemingly natural but undesirable conditions may also be due to undiscovered microbial agents, especially since bacteria outnumber tissue cells in our highly infested 20th century bodies by more than 10:1.

Unwanted syndromes

Syndromes are groups of related symptoms and signs of disordered function that define a disease whose cause remains unknown, that is, idiopathic. A good example is irritable bowel syndrome (IBS), which affects up to 20% of the adult U.S. population and includes symptoms of abdominal distention and pain, with more frequent and looser stools. Many are unaware they are afflicted. In 1998 there was no known cause or simple complete treatment for this still "natural" disease. Even more mysterious than IBS is our general activity level—some people seem to have high-energy personalities, while others have more phlegmatic low-energy personalities. Either may be regarded as "natural," but nanomedicine can probably bring this ill-defined neurophysiological variable under human control. The need to sleep is another imperfectly understood syndrome. It is experienced by everyone and thus was universally regarded as "natural" in the 20th century. Physiological short sleepers were unusual, insomnia or asomnia was thought of as an abnormal state, and there were a few anecdotal but medically undocumented instances of total nonsomnia, such as the celebrated case of Al Herpin.

Psychological traits

Psychological traits which, if identified by a patient as undesirable, might be subject to genetic or physiological modification could include: sexual preference (6-10% of the adult population is homosexual); shyness or boldness; acquisitive or altruistic propensity; misanthropy or philanthropy; theistic or atheistic orientation; loquacity or dourness; childhood imprinting; criminal propensity (up to 1-5% of the population); various recognized personality disorders that affect ~10% of the population such as antisocial, paranoid, schizotypal, histrionic, narcissistic, avoidant, dependent, obsessive-compulsive, and passive-aggressive disorders; panic attacks (experienced at least once by ~33% of all adults each year); and phobic disorders such as social phobias (~13% of the population), specific phobias including fear of large animals (*zoophobia*), snakes (*ophidiophobia*), spiders (*arachnophobia*), needles (*belonephobia*, ~10%), the dark (*noctiphobia* or *scotophobia*), or strangers (*xenophobia*) (total ~5.7%), the fear of blood or *hemophobia* (~5%), *agoraphobia* (~2.8%), and other unusual phobias such as the fears of certain colours (*chromophobia*), daylight (*phengophobia*), girls (*parthenophobia*), men (*androphobia*), stars in the sky (*siderophobia*), the number thirteen (*triskaidekaphobia*), and even the fear of developing a phobia (*phobophobia*).

The above sampling of minor afflictions, almost all considered "natural" in 1998, may come to be regarded as commonplace correctable medical conditions in the nanomedical era. By the time such petty annoyances are deemed worthy of immediate treatment, biotechnology and nanomedicine already will have defeated the most fearsome illnesses of the late 20th century and will have moved on to other challenges. Naturophiles may dissent, but the emerging trend from medical biotechnology is to characterize health, not as a static standard, but rather as a condition defined by the lives that people want to lead. Affirming the volitional normative model of disease, Virginia Postrel concludes:

"Different goals will produce different choices about trade-offs and standards. What makes a condition unhealthy is not that it is unnatural but that it interferes with human purposes. Revering nature [would mean] sacrificing the purposes of individuals to preserve the world as given. It [would require] that we force people to live with biological conditions that trouble them, whether diseases such as cystic fibrosis or schizophrenia, disabilities such as myopia or crooked teeth, or simply less beauty, intelligence, happiness, or grace than could be achieved through artifice. In a world where it's no big deal to take hormone therapy, Viagra, or Prozac, to have a face lift, or to know a child's sex before birth, a world in which even such radical interventions as sex-change operations and heart transplants have failed to turn society upside down, it is extremely difficult to argue that medical innovations are dangerous simply because they fool Mother Nature."

2

NANOETHICS

Nanotechnology is a rapidly developing field of technology that seems to have the potential of great upsides and excessive downsides. Thus far, there has been a strong tendency in the debate on nanotechnology to focus on either the first or the latter. Accordingly, assessments of nanotechnology tend to radically diverge. On the one hand, optimistic visionaries promise truly utopian states of affairs, e.g. solving the problem of hunger in the world or expanding our maximum life span. On the other hand, pessimistic thinkers draw worst-case scenarios in which, for instance, nanotechnology has exceptionally disruptive effects on societies or swarms of nanoassemblers devour the whole biosphere. The utopian views follow from one-sidedly focusing on the potential benefits of nanotechnology, whereas the apocalyptic perspective results from giving exclusive attention to worst-case scenarios.

These radically diverging assessments that have thus far dominated the debate on nanotechnology seem to a lack common ground. This situation holds the risk of conflicts and unwanted backlashes. Hence, the present state of the debate on nanotechnology calls for the development of a more balanced ethical view. This contribution will first briefly describe the field of nanotechnology. Next, the present state of the ethical debate of this field will be described as overshadowed by utopian dreams and *apocalyptic nightmares*. Finally, a method will be introduced to develop more balanced ethical views on nanotechnology. Thus, the focus of this chapter is on the *methodology* and not on *normative analysis*.

TERMINOLOGY AND BASIC IDEA

Nanotechnology is a rapidly developing new field of research. In the literature, both a fairly broad as well as a rather narrow concept of nanotechnology are employed. The first signifies any technology smaller than microtechnology. In contrast, the latter stands for the technology to program and manipulate matter with molecular precision and to scale it to three-dimensional products of arbitrary size.

The basic idea of nanotechnology, used in the narrow sense of the word, is to employ individual atoms and molecules to construct functional structures. In his lecture "There is plenty of room at the bottom", at the annual meeting of the *American Physical Society* at the *California Institute of Technology*, the famous physician Richard Feynman already speculated on radical forms of miniaturization. His reflections, however, were not taken very seriously at first. Both practically and theoretically, significant progress in the field of nanotechnology started only in the eighties.

Practical Development

A significant practical development was the fast progress in microscopy. At the beginning of the eighties, Gerd Binnig and Heinrich Rohrer, working at the IBM research laboratory in Zurich, developed *scanning tunneling microscopy*. This new technique can provide an image of the atomic arrangement of a metal or a semiconductor surface. Thus, using this new technique Binnig and Rohrer could, for the first time ever, "*map*" the arrangement of individual atoms of metals and semiconductors. In 1986 Binnig and Rohrer received the Nobel Prize for their achievements.

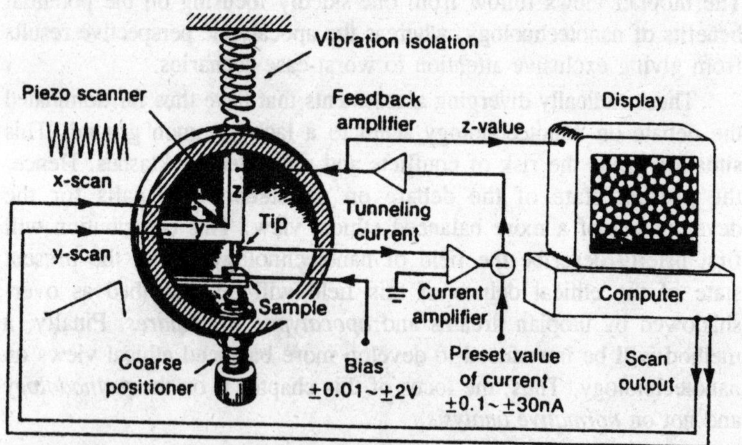

Fig. 2.1. Scanning tunneling microscope.

The *atomic force microscope* represents a further development in microscopy. It enables images of materials inaccessible to the scanning tunneling microscope, for instance insulators, organic materials, biological macromolecules, polymers, ceramics and glasses.

Finally, it was discovered that scanning tunneling microscopes could also be used to manipulate nanoscale objects. For instance, in 1990, Eigler and Schweizer, two IBM scientists, were able to shape the three initials of the logo of their employer from 35 individual xenon atoms. It is expected that further developments in microscopy will be highly significant for nanotechnology.

Theoretical Development

A significant theoretical development in the eighties was the research done by Eric Drexler (1981 & 1986). His work provided a theoretical basis for the field. Besides, it brought *nanotechnology* to the attention of a broader public. Drexler argues that the laws of physics do not forbid the possibility of pre-programmed maneuvering and goal directed management of individual molecules. In addition, he elaborates on the future development of technical means and methods to arrange matter at the nanoscale. Finally, Drexler also discusses the various fields of application for future nanomachines.

In discussing future nanotechnological manufacturing, Drexler comes up with the idea of the assembler: „a molecular machine that can be programmed to build virtually any molecular structure or device from simpler chemical building blocks". This nanoscale construction device can position molecules in every which way, thereby facilitating, for example, chemical reactions. Through precise sequences of manipulations, a computer-steered assembler could thus—molecule by molecule—assemble any chemically stabile structure that it has been programmed to construct.

According to Drexler, the development of universally applicable assemblers is essential for the further development of nanotechnology. Assemblers could obviously also be programmed to replicate themselves. From a commercial point of view, this would have the interesting advantage of being able to manufacture products in a feasible time frame. After all, if accomplished by only one assembler, constructing a car in a molecule-by-molecule way could take ages. If, on the other hand, millions or billions of assemblers could work together, things would look far more optimistic.

State of the Field at Present

In the meantime, the field of nanotechnology has developed considerably. Current research is not exclusively focused on achieving assemblers. Instead, research is directed towards the production of a wide array of different nanoscale structures. The fabrication techniques of these structures can be divided into two approaches: "top-down" and "bottom-up".

The *top-down techniques* that are used to manufacture nanoscale structures are mostly extensions of methods already employed in small-scale assembly at the micron scale, for example, *photolithography*. By further *miniaturization*, the *nanodimension* is entered. In this way, further miniaturization of microelectronics could result in nanoelectronics.

Bottom-up fabrication methods for manufacture are studied within synthetic chemistry, which is, almost by definition, the science of producing nanoscale structures. They are also inspired by phenomena such as crystal growth and self-assembly. In a certain way, many bottom-up methods try to imitate regularly occurring processes in nature. Living nature, for example, constantly shapes complex macroscopic structures from individual biomolecular elements.

It is widely expected that, if nanotechnology were to be further developed, a whole range of products could be produced more rapidly, cheaper and better. After all, manufacture would be done in a molecule-by-molecule way, meaning that all features of the product and all aspects of production could be programmed and would be under control in ways that are unknown up to now. Conformingly, many countries have set up programs to financially support further development of nanotechnology. Moreover, investing in the further development of specific nanotechnological projects is also rapidly getting more interesting for private companies.

Utopian Dreams and Apocalyptic Nightmares

The present state of ethical debate about nanotechnology seems to be dominated by utopian dreams and apocalyptic nightmares.

Utopian Dreams

During the 17th century, the idea was developed that *utopian ideals*, such as control of the natural environment, a perfect society, life without disease and pain, prolongation of life as well as enhancement of man and his characteristics could be achieved through the further development of science. All that was needed for that would be to organize science in a correct way and to work with effective methods. This modern idea became a source of great enthusiasm.

Many contemporary views on nanotechnology can be placed in this tradition of scientific utopianism, which started in the 17th century.

Over and over again, it is argued that, if only nanotechnology were to be fully developed, a large part of the world's current problems would be solved and a whole array of ideals would be achieved. A few examples of the utopian perspectives of nanotechnology are given.

To begin with, molecule-by-molecule manufacturing would be self-sufficient and dirt free. After all, molecular manufacturing techniques would not result in any chemical pollution whatsoever. Leftover molecules would be recycled. What is more, molecular manufacturing would enable environmental restoration at the molecular level. Unwanted chemicals could be detected and inactivated. Hence, we will be able to reverse existing environmental degradation.

Next, molecule-by-molecule manufacturing could create unprecedented objects and materials. For example, new nanotechnologically manufactured strong lightweight materials will enable easier access to space and space resources. *Nanoelectronics* could come up with computer chips that would be billions of times faster as a result of the smaller components.

Using *molecular manufacturing techniques*, we will be able to produce inexpensive high-quality products. For example, storage batteries, processors, personal computers, lap tops, cell phones and display devices could become strikingly inexpensive.

Molecular manufacturing could also be used to fabricate food rather than growing it. After all, food is simply a combination of molecules in certain configurations. Hence, the problem of hunger could be effectively solved by efficient molecule-by-molecule mass production of food.

Finally, it is also in medicine that nanotechnology is said to work miracles. There are numerous astounding promises in this context. Molecular manufacturing will provide low-priced and superior equipment for medical research and health care. These improved tools would be available far and wide. *Medical nanomachines* will be programmed to travel through our bloodstream to clean out fatty deposits. Hence, they would reduce the probability of cardio-vascular diseases. Medical diagnosis and drug-delivery will be transformed. Moreover, preventive medicine will be greatly improved by having nano-robots within our bodies that could provide a defense against invading viruses. Thus, nanotechnology has been hailed as the solution to many medical problems.

It is even expected that it will contribute to the enhancement of man. Not only will it be possible to overcome contemporary diseases, pain and other unpleasant bodily symptoms. Over and above, nanotechnology will enable us to enhance all our human capabilities and properties.

With regard to the enhancement of the human body, it is expected that nanotechnology will enable the construction of stronger and enhanced tissues and organs. For instance, cells specific to certain tissues or organs could be reconstructed and be made immune against all known pathogens, thereby making our present immune system obsolete.

The *cryonics community* has also enthusiastically embraced nanotechnology. Cryonics means freezing people who have been declared legally dead and waiting until technology is advanced enough to reverse cause of death as well as the freezing damage. Here, nanotechnology is expected to produce real miracles in reversing all the adverse configurations of molecules in the frozen organism after it has been thawed out.

Also, nanotechnology would enable an almost infinite improvement of our mental capacities. It would, for example, be possible to enhance our memory as well as all our data processing capacities. However, with regard to enhancing the human mind the scenario of '*uploading*' is the *non plus ultra*. Uploading involves transferring the contents of the human brain to a computer. Special nanomachines would scan the brain atom-by-atom. Next, the neural networks of the brain would be implemented on a computer.

Nanotechnology is finally also expected to lead to social advances. Freitas (1998b) thinks that the huge achievements of nanotechnology, especially nanomedicine, will make people more content and peaceful. It will be a great deal easier to live together in ideal harmony with perfect bodies and flawlessly functioning brains.

Apocalyptic Nightmares

Besides utopian outlooks, catastrophic scenarios have also heavily influenced the debate on nanotechnology. Severe disruption of many different aspects of society and politics is one of the nightmare scenarios that have been sketched in connection with the further development of nanotechnology. For example, rapid developments in molecular manufacturing and the concomitant inexpensive manufacturing could cause severe economic disruption. The economic upheaval could involve the sudden abundance of low-priced products, rapidly changing employment patterns (e.g. unexpected redundancy of a variety of jobs) and the problem of copying of designs.

Moreover, molecular manufacturing might also invite premeditated misuse in warfare or terrorism. First, all kinds of conventional weapons could be constructed more rapidly. Next, new *nanoweaponry* and could be made in huge numbers, low-priced, extremely powerful. Hence, traditional arms control would be far more difficult and competing states could enter a troublesome and unsound arms race that would be extremely difficult to end.

Also, infinitesimally small surveillance devices such as *nanoscale* tracking devices, *nanosensors*, *nanocameras* and *nanomicrophones* could enable dictatorial observation and control of subjects in a way that is totally unprecedented. Nanotechnology would enable total surveillance of entire civilian populations without them even noticing it.

Nanotechnology could also have the potential to cause extensive environmental damage. For example, destructive nanomachines might enter the food chain thereby disturbing entire ecological systems or nanomaterials such as *nanoparticles* could escape into the air and turn out to pose asbestos-like health treats.

Some of the most serious risks of nanotechnology have been brought to the attention of the public by Bill Joy, co-founder and scientific leader of *Sun Microsystems*. Joy is especially worried about the research with regard to assemblers. After all, these nanomachines will have the worrisome capacity of self-replication. Without this kind of assemblers it is hardly imaginable how molecular manufacturing could ever become practically feasible. After all, without the ability of self-replication, all assemblers needed for nanotechnological production would have to be built one by one, which would definitely be too expensive. In this case, however, practical feasibility and commercial viability would involve grave dangers. Technical faults, for example problems with the software of the onboard computer of an assembler, could cause unbridled self-replication. In that case, since the newly produced assemblers would also start replicating themselves, the total number of assemblers would grow exponentially. If these uncontrolled assemblers used a wide variety of raw materials as resources for self-replication, they could devour the whole biosphere in an amazingly short while. The biosphere would, so to speak, be transformed into gray goo—hence, the terminology of *gray-goo* scenario.

Another danger has to do with the fact that many private companies will try to develop and produce assemblers. After all, the perspective to put them on to the market will seem lucrative for many. Hence, it will be difficult for central governments to retain control over the development of assemblers. Therefore, there will always be the danger of abuse of assemblers by criminals and terrorists, for example, to develop weapons of mass destruction. For example, one could construct special nanomachines to destroy computer systems or certain resources of the opponent. Of course, nanomachines could also be designed to attack the enemy directly. This apocalyptic scenario, in which *destructive nanomachines* are being used for warfare or terrorist purposes, has also been called the *black-goo*-scenario.

Toward a More Balanced Ethical View

The dominance of the drastic opposition of utopian dreams and apocalyptic nightmares in the debate on the future perspectives of nanotechnology holds the risk of undesirable conflicts and unnecessary backlashes. Hence, the present state of debate on nanotechnology calls for the development of more balanced ethical views. In response to this important challenge, a six-step method is here presented. With the help of this method, a fine-grained and rational assessment can be made as to the ethical desirability of further developing research in a specific field of nanotechnology.

Specific Field of Nanotechnology

Up to now, broad and sweeping statements about nanotechnology as such have dominated the debate. However, for an ethical analysis to be sound and discriminating, it should be focused more specifically on a particular field of nanotechnological research, instead of generalizing in an all-encompassing way. Thus, more detailed and better-informed ethical research is needed. Nanotechnology is by no means one single effort. Rather, it is a complex of countless different projects with a huge variety of goals. Hence, different fields of nanotechnological research can be distinguished that will not necessarily demonstrate identical or even similar ethical aspects, for example: (i) materials and manufacturing, (ii) nanoelectronics and computer technology, (iii) medicine and health, (iv) aeronautics and space exploration, (v) environment and energy, (vi) biotechnology and agriculture and (vii) security. Evidently, the ethical assessments of developments in these fields are likely to differ as the objectives and the ethical problems encountered will be different.

Objectives of Specific Field Nanotechnology

Having specified the nanotechnological field to ethically be assessed, the next step is to focus on the objectives that the research in that field aims to achieve. Ideally, we should first discuss our needs, fundamental purposes and social ends in order to move on to make choices about ways of achieving these goals, for example by further developing certain fields of technological research. However, this is not always the case. What is more, technology often seems to develop in a seemingly autonomous way. However, it remains imperative to reflect about the goals that we try to achieve in developing certain technologies. Technology development should be directed towards a good or an end. Otherwise, technology is developed for its own sake, isolated from any human good. This would obviously be undesirable. After all, science and technology should serve man and not the other way round.

Objectives Ethics

After having specified the nanotechnological field to be assessed as well as having detected the objectives of that field, it should be asked whether these objectives are ethically desirable. After all, if the objectives that are aimed for with the further development of a certain research area are ethically undesirable, it does not make sense to continue this research from an ethical point of view.

Development of the Field

If the objectives are ethically desirable but not achievable or very unlikely to be achieved by advances in a specified field of research, it seems pointless to push the research forward. Of course, it is not possible to predict advances in science or technology in detail and with certainty. Nevertheless, in many cases it will be possible to give a broad assessment of the probability of achieving certain objectives in a specified field of research on the basis of the corpus of existing scientific literature.

Ethical Problems

The next step involves determining the ethical problems that are connected with the further developments in a specified nanotechnological field. *Ethical problems* connected with further development of nanoresearch are not necessarily the same in different fields of nanotechnological inquiry. For example, present research on nanotechnologically manufactured coatings of prostheses to improve their biocompatibility can hardly be said to pose the same ethical questions as, for instance, projects focused on producing perfect mosquito nets to reduce the problem of infectious disease or research on new nanochips that can read individual genomes in a feasible timeframe. Thus far, unfortunately, ethical problems have been mostly discussed on a very general level. As a rule, problems have not been linked to specific fields of nanotechnological enquiry. In order to give an idea of the ethical problems that have been discussed in debate on nanotechnology up to now, the most important problems are here presented without pretending to give a complete sketch.

Risks

Up to now, the ethical debate on nanotechnology has been dominated by discussions about risks. A few examples are risks of disruption of the basis of economies, environmental damage, an unstable arms race, the gray goo scenario, the black goo scenario etc. Against this backdrop, the environmental pressure group ETC has proposed that governments worldwide proclaim an instantaneous moratorium on commercial fabrication of new nanomaterials. Moreover, they want to start a transparent and global process of assessment of the various implications (for example for the environment, society and health) of the technology. Other less radical authors have called for an early and honest assessment of all the risks and unintended consequences of nanotechnology.

Equity

From a global justice point of view, the risk that the rise of nanotechnology will only be advantageous for rich countries, leaving developing countries behind, should be avoided. Therefore, it is imperative that a strategy be developed as to how the development of nanotechnology can be organized to both improve the lives of people in industrial states as well as advance living standards of those in developing countries. Also within individual countries and societies a sharp nano-divide between the well off and the underprivileged ought to be prevented. It would clearly be problematic from the viewpoint of equity, if further development of nanotechnology would disproportionately benefit the higher echelon of society. In that case existing gap between haves and have-nots would only widen.

Privacy

The extreme miniaturization that nanotechnology facilitates would be enormously advantageous for further developing espionage and surveillance gadgets. It is not hard to imagine infinitesimally small tracking devices, nanocameras and nanomicrophones registering virtually everything we do without us even noticing it. Evidently, this idea poses the question of whether, and if so how, we will be able to defend our privacy.

Playing God

From a religious point of view, it could be argued that fully advanced molecular manufacturing would demonstrate a problematic attitude towards God's creation. After all, it would involve a fairly fundamental reshaping of creation in order to comply with human ideas and directives. Taking God's creation apart in the most extreme form and putting it together in a molecule-by-molecule way is likely to offend those who believe that the structure of creation is already pervaded by divine rationality and heavenly benevolence.

Approach to nature

In a way that would not necessarily have to be based on religious premises, a similar argument could be constructed around the attitude to nature that is involved in molecular manufacturing. Constructing our surroundings in a molecule-by-molecule way demonstrates an invasive and instrumental approach to nature that could be seen as conflicting with the idea of respect for nature. The idea that nature has finality and some kind of moral meaning is a very old tradition in western philosophy. It can be found already in the philosophy of Plato, Aristotle, the Stoics, the Neoplatonists and Thomas Aquinas. More recently Goethe, in his theory of colours, criticized Newton for having forced light through prisms, thereby demonstrating a wrong approach to nature and thus coming up with biased results as to the real essence of light.

Surmountable Ethical Problems

After having determined the ethical problems that are connected with further developments in a specified nanotechnological field, analysis should decide whether they can be dealt with or not. After all, if it happens that the ethical problems are not surmountable, it seems ethically objectionable to favour further developments of the research.

To exemplify this last step, an example is given that concerns risk management. To reduce the risks of the further development of techniques of molecular manufacturing using assemblers the Foresight Institute has issued the so-called "*Foresight Guidelines on Molecular Nanotechnology*". These guidelines contain a set of "*Development Principles*" as well as "Specific Design Guidelines". Among the first are prohibitions such as the statement that artificial replicators must not be capable of replication in natural environments or directives reminding developers that they should attempt to consider systematically the environmental consequences of the technology and limit them to intended effects. Among the "*Specific Design Guidelines*" are directives focussing on ways to device designs to prevent self-replicating devices running amok such as complete dependence on artificial "*vitamins*" not available in any natural environment or programming termination dates into devices.

Conclusion

Hardly ever has there been such a discrepancy between opposing evaluative judgments as can be observed in the debate on nanotechnology. However, both the radically optimistic as well as the thoroughly pessimistic ideas seem to be founded on a somewhat one-sided and narrow approach in ethical assessment. Nanotechnology is a complex endeavor that consists of many different projects with a huge variety of goals. Hence, different fields of nanotechnological research should be distinguished. After all, these fields will not necessarily demonstrate identical or even similar ethical aspects. Up to now, most of the debate has been too emotional, too general and quite decontextualized.

In order to improve ethical assessments of nanotechnology, a six-step method has been presented. Use of this method results in a rational and systematic assessment of the ethical desirability of further development of a specific field of nanoresearch. Due to the complexity of nanotechnology, it will be necessary to create interdisciplinary groups for ethics research. These, in turn, would need adequate funding. Therefore, it would be a good idea to devote a certain percentage of the government spending for nanotechnology to ethical research such as we have done in earlier attempts to address the ethical, legal and social implications of the human genome research.

3

DIGITAL TECHNOLOGY

Charles Babbage is famous today for almost inventing the computer. Specifically he invented it, and almost got it built, but not quite, and nobody bothered with it again for a century. Once people finally did build working calculating machines, computers took off like crazy. The twentieth century has seen a trend in computer power that is remarkable in the history of technology.

Just like physical power engines, computing engines have stayed on an exponentially improving track through a series of changes in the underlying technology. Computers started with mechanical shafts and gears, as in Babbage's design, followed by electromagnetic relays, vacuum tubes, discrete transistors, integrated circuits, and a series of order-of-magnitude leaps in the size of the circuit that was integrated.

Just as with the steam engines, this leads us to ask whether our computers might be approaching some theoretical limit. The answer turns out to be yes. In fact, it is the same thermodynamic limit Carnot discovered, but in a different disguise.

Modern-day *thermodynamics* talks about such questions in terms of a property called *entropy*. Entropy has found its way into the general vocabulary and is widely misused and misunderstood; but in thermodynamics it has a precise and unambiguous meaning. For example, a specific quantity of a working fluid in a heat engine has an entropy that depends on its temperature and volume. Entropy is added when heat is added, in a quantity proportional to the temperature. In an adiabatic process, entropy remains the same. But the bottom line is that taking the sum of entropy for all the parts of the engine, including the hot bucket and the cold bucket if the process is reversible, the entropy stays the same; otherwise, it must increase.

In 1948 a communications engineer at the fabled Bell Labs published a paper in two parts in the July and October numbers of the *Bell System Technical Journal*. His name was Claude Shannon, and this paper was titled "A Mathematical Theory of Communication." It was among the greatest pieces of scientific analysis of the twentieth century, easily rivaling Carnot's establishment of thermodynamics. The field Shannon invented is now called information theory. Among other things, he established the "*hit*" as the basic unit of information. He showed that any information-bearing signal could be converted to bits, transmitted, and converted back again. And what's more, he proved that this was the most efficient way of transmitting information. Now-a-days anything can be converted into bits for granted nowadays, but it was an unheard—of idea at the time. (By the way, after the breakup of the Bell system, the part of Bell Labs that remained with AT&T was named Shannon Labs in his honour.)

One of the things that Shannon did in information theory was notice that the mathematics of what happened in information channels looked a lot like the mathematics of thermodynamics, so much so that he called one of the measures information entropy. Shannon was smart: it turns out that information is exactly what thermodynamic entropy actually measures.

One of the most puzzling conundrums of science over the past 150 years or so is the notion of maxwell's Demon. James Clerk Maxwell, the same one who integrated the physics of electricity and magnetism into the elegant set of equations that bears his name, also wrote a book about thermodynamics, and in it he describes a thought experiment:

One of the best established facts in thermodynamics is that it is impossible in a system enclosed in an envelope which permits neither change of volume nor passage of heat, and in which both the temperature and the pressure are everywhere the same, to produce any inequality of temperature or of pressure without the expenditure of work. This is the second law of thermodynamics, and it is undoubtedly true as long as we can deal with bodies only in mass, and have no power of perceiving or handling the separate molecules of which they are made up. But if we conceive a being whose faculties are so sharpened that he can follow every molecule in its course, such a being, whose attributes are still as essentially finite as our own, would be able to do what is at present impossible to us. For we have seen that the molecules in a vessel full of air at uniform temperature are moving with velocities by no means uniform, though the mean velocity of any great number of them, arbitrarily selected, is almost exactly uniform. Now let us suppose that such a vessel is divided into two portions, A and B, by a division in which there is a small hole, and that a being, who can see the individual molecules, opens and closes this hole, so as to allow only the swifter molecules to pass from A to B, and only the slower ones to pass from B to A. He will thus, without expenditure of work, raise the temperature of B and lower that of A, in contradiction to the second law of thermodynamics.

The reason Maxwell's Demon posed a serious dilemma is that he appeared to be able to break the *second law of thermodynamics*. Maxwell gives a good account of it in the passage quoted; today we would be more likely to say that he lowers the entropy in his "vessel" but does not raise it anywhere else.

Maxwell's Demon offended three generations of physicists before his facade was cracked by information theory. The first reasonable assault was to note that information theory could he used to characterize the "*channel*" of information flow, which involved his seeing the molecules, and that the information entropy he gained by recognizing fast and slow molecules just balanced the thermodynamic entropy he "*magically*" removed from the system. This was a lot better than nothing, but it was not airtight—what if the demon used something other than sight to recognize atoms?

Suppose, for example, you gave the demon a cheat sheet that tells him what times to hold the door open and when to close it. It could be a player piano-roll, so that the demon is just a simple mechanism. Nothing supernatural at all. And yet it would work. Now the demon has no contact with the outside world at all. You, of course, had to obtain the information from the molecules when you made up the sheet, but the demon is completely cut off during the process.

Here is the fix: noting that entropy is a measure of information, the cheat sheet has enough information on it to square up the apparent violation of the second law. The demon can separate hot and cold without violating thermodynamics, but he cannot erase the sheet. With the sheet still there, the whole process is reversible; in theory, anyway, you can run all the atoms backward and have him read the sheet back ward, putting all the molecules back where they came from.

In other words, the entropy represented by the information on the cheat sheet is just as real as the classical physical form in the vessel—so as long as the demon does not erase the sheet, he has not really reduced the entropy.

What the second law says to modern, information-theory-aware physics, then, is that information cannot be erased. It even tells you exactly how much heat at any given temperature is equivalent to a bit of information. It turns out to be in the ball park of the heat energy of a single atom at that temperature. A simple way to understand why heat can be used to store information is that hotter atoms move faster, and thus require bigger numbers—more bits—to describe their speeds. Thus more information has been "*stored in*" an atom with a higher speed. You can also increase entropy by making the system physically larger. This stores more information by allowing more bits in the description of an atom's position.

So here we sit with computers that erase bits by the millions. The second law now tells us that this comes with an unavoidable price: you may think you are erasing bits, but at the molecular level you are just converting them to information in heat form.

For today's computers, this is not a problem: the amount of heat they dissipate is so much bigger than the thermodynamic limit that the difference is not noticeable. But the computers of tomorrow, with switching elements of molecular size, will find that the limit is a problem. Not one that cannot be gotten around, just one that will need to be addressed. We can first build computers to erase as few bits as possible and then simply put cooling systems on to them.

Surprisingly enough, the "*cannot erase bits*" formulation of the second law affects machines besides computers. A simple example is a common door latch for an interior door. The bolt protruding from the door is beveled and spring loaded. When you push the door closed, the force of closing pushes the bolt back, and then it pops into place in the strike plate hole. The process is irreversible—you cannot open the door simply by pulling; you have to turn the knob.

That *irreversibility* erases information. If the door were a swinging door with no friction, you could tell how fast the door had been swinging a minute ago by how fast it's swinging now. The latch erases that: once it's stopped, you cannot tell what it was doing before. Notice that friction erases the same information; but we already knew that friction dissipates heat.

If you could build a one-way mechanism like a door-watch at the very lowest of molecular levels, you could convert the random vibrations that constitute heat into coordinated motion and force, that is, useful work. But the second law, in its information theory form, stands in your way. For every bit your mechanism forgets, it has to dissipate just exactly as much heat as the amount of energy it could have gained from heat by its irreversible action.

You could, of course, build a gadget that looked like a door latch at the molecular scale. But if the spring were weak enough that it could be closed by thermal vibrations, the very same thermal vibrations would happily open it, too: it would not be irreversible.

Bottom line nanomachines cannot act as Maxwell's Demon to break the second law of thermodynamics. They cannot turn raw heat into useful work. Of course, they can turn a temperature difference into work, just as a macroscopic heat engine can.

As an aside, it seems likely that the mechanical losses such as friction can be kept to a minimum, so that relatively small temperature differences can be used effectively. For example, a human body might produce 200 watts of heat at about 100 degrees Fahrenheit and dissipate it into a room temperature environment of 70. Thermodynamics says that only about 5% of that can be recovered as useful work—10 watts. It's hard to imagine a macroscopic machine that could harvest that—you can lose 5% of your power in each gear or bearing`in small mechanicals. Nanomachines might be able to do it, though.

Great Revolution

It is *ironic* that munch of what we know of the theories of Democritus comes to us through the works of *Aristotle*, who disagreed vehemently with his physical theories. Although not the first, Democritus was the most thoroughgoing, systematic, and clearly the foremost proponent in his day of the notion that matter was composed of atoms.

Chemists in the nineteenth century, notably John Dalton., revived the idea because of an interesting property that kept cropping up in chemical reactions. If you ignite 2 grams of hydrogen in 16 grains of oxygen, you'll get 18 grains of water. But if you use 3 grams of hydrogen, you'll get the same amount of water and have a gram of hydrogen left over. Chemicals react in definite proportions. Chemists were forced to the conclusion that matter, from the densest metals to the most tenuous of gasses, came in discrete, indivisible chunks.

It took the physicists some time to catch up. It wasn't until the twentieth century that atoms were fully accepted physical theory. There was a time when some physicists believed that discrete electrons were embedded in otherwise continuous matter, like raisins in tapioca pudding. However, with the help of phenomena like *Brownian motion*, seeing tiny spores dance around inexplicably in water under a microscope, physicists came around and by the early 1900s were busy figuring out how atoms work, untimately giving us quantum mechanics. It's another irony of science that Einstein never won a Nobel Prize for relativity; his prize, in 1921, was for the explanation of the *photoelectric effect* in establishing quantum theory—that light also comes in discrete dunks.

So by 1948 when Shannon's information theory came out, it was well accepted that matter came in discrete chunks, and that the chunks were of a small number of types. Shannon then proceeded to show that information, too, even continuous information like the waveforms of sound, was ultimately decomposable to hits. Mind as well as matter, it seemed, was digital.

The biologists weren't far behind. Within a decade, life itself had fallen to the digital paradigm. In 1950 DNA was a chemical. In 1960 it was a reel of computer tape. Over the second half of the century, biologists worked out how things like ribosome's were numerically controlled machine tools at the molecular scale.

In 2003 digital cameras outsold analog, film—using ones. By now, there's a good chance that most of the pictures you see and most of the music you hear is captured, processed, and presented digitally. The text you are reading certainly is being produced and manipulated digitally; all except for the final inked page of the physical volume you hold. The digitization of information has allowed the complete and utter transformation of our ability to handle, manipulate, and trans form it. It is the digitization of matter that allows life to exist at all. And the control of matter in a digital way is exactly what nanotechnology is all about.

Why; after all, should we bother to build machines at the molecular scale? Large machines work just fine, and they are a lot easier to work with. And if there is some great advantage in building small, why stop at the nanometer scale? Why not *picotechnology*, *femtotechnology*, all the way down to *zeptotechnology*?

The answer to both questions is that the atomic scale is where matter is digital. Atoms come in a small number of discrete types and form bonds in a small number of fairly simple ways. Atoms don't wear out, and atoms of the same kind are all exactly the same. Nature has a very good quality control. To make a part that is just the same as another in a molecular machine, you take the same kinds of atoms and connect them with the same pattern of bonds. Then, in a very strong and useful sense, the new part is exactly the same as the old.

In macroscopic machines, it has long been understood that inter changable parts are a major advantage. However, no two macroscopic parts are ever actually the same. There is always a tolerance—the engineer's term for a margin for error in size. Not only does this mean that two parts won't act exactly the same, but that no part is exactly what was designed. Thus machines are not as smooth, quiet, or efficient as they could be.

A lot of properties of everyday matter can vary with age and wear. An atomically precise part cannot wear: either all the atoms and bonds are there that are supposed to be or it's broken. Covalent bonds between atoms in molecules don't fatigue. The same thing is what makes digital computers so much more reliable, and cheaper, than analog ones. A bit is either right or wrong, never 97.4% right. A computer can do trillions of operations in a row without an error. That's virtually unheard of from an analog machine.

We understand quite a bit about taking advantage of digital properties from computers. We can use many of the same techniques in physical machines with nanotechnology. When nanotechnology makes machines cheaper and more reliable, it will be because of the digital nature of matter at the atomic scale.

NANOCOMPUTERS

Nanoelectronics is one of the leading areas of nanotechnology research today. In fact, all the components necessary to make a *nanocomputer*—molecular switches and nanowires—have been demonstrated. "*Demonstrated*" in many cases means that a lab technician shifted through chemically generated molecules for months with very expensive equipment and found a few that worked as desired. In other words, we know they exist—which is great for a scientist, but not so good for an engineer. There is as yet no way to put them together in large, complex circuits, much less to do so economically. Once we have the nano robot arms, though, it will be straightforward.

Most electronic switches in use today are transistors. Various molecules have been shown to function as transistors. Once we have the ability to place atoms in arbitrary arrangements, the number of different kinds of known transistors will in all likelihood go through the roof.

Other forms of switches will also work at the *nanoscale*, and may have uses. *Electrostatic* relays would work. They could have fast switching times compared to today's electronics (but not to *nanoelectronics*). They might find a use in power applications—turning motors and lights on and off. Note that at the nanoscale, the distinction between a transistor and a relay can get fairly fuzzy. Transistors operate by having a voltage change cause a change in the conductivity of a material. In appropriately structured molecules, conductivity can be altered by moving other molecules nearby or by bending them—phenomena that seem partway between moving a physical switch and alter a property by electric fields. *Molecular switches* have been demonstrated in the lab using a number of different such techniques.

Most *nanomachines* will have *nanoelectronics* built in, to control all those motors, and computers to control the controllers.

In current-day computer chips, getting all the transistors you need is not the problem: the problem is the wires. There just is not room enough to connect everything you want with everything else. And wires have increased resistance as they get smaller.

You want wires with low resistance for two reasons. In power applications, from transmitting power to your house across the country to providing power to all those little motors that move your robot, resistance simply wastes power as heat. Lower resistance is just more efficient. That is true in computers, too, but there is another reason. Resistance makes computers slower.

Suppose you have a computer with a 1-gigahertz clock speed. That means that the clock signal rises and falls once a *nanosecond*—the time it takes light to travel about a foot. But the signals inside the chip are traveling, much less than a foot. What's happening in the computer is that signals move from switch to switch by electrons filling up or draining out a wire and the part of the switch connected to it. The fullness of a wire with electrons is its voltage. Switches do not switch until the voltage changes.

Think of a *switch*, a *transistor*, for example, as a valve turned on or off by the weight of a bucket. The bucket is filled or emptied by hoses that are controlled by other valves. The crucial time for switching is not how long it takes water to start coming into the next bucket; it's how long it takes to fill (or empty) the bucket. With nanoelectronics we can make smaller buckets; the question is how big can we make the hoses?

Mechanical Nanocomputer

Electronic nanocomputers will be very fast and small compared with current electronics, but mechanical nanocomputers might be smaller still. The reason is that electrons are hard to nail down and have interesting quantum properties as the scale approaches molecular size. For example, an electron siting in a hole it doesn't have the energy to get out of can suddenly be found in another hole some distance away. This ability to jump without having been in the intervening space is called *tunneling*.

The probability an electron will tunnel depends very strongly on the distance it jumps. This property is used in one of the very first scanning probes, the *scanning tunneling microscope* (STM). In the STM, electrons tunnel from the probe tip to the object being probed, and the distance can be measured quite accurately because the current changes drastically as the distance changes minimally.

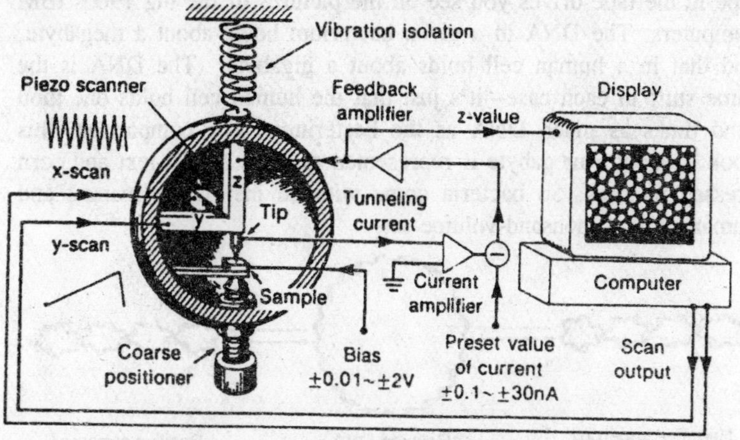

Fig. 3.1. Scanning tunneling microscope.

In nanocomputers or other circuitry, we are going to have to keep wires three *nanometers* or so apart or else electrons will tunnel from one to the other. In other words, at the nanoscale, wires don't have to touch to cause a short circuit. So nanocomputers using conventional wires, switches, voltage, and current will have to be bigger than the sizes of their components would suggest.

We could make *mechanical nanocomputers* that used sound conducted along strings, like a tin can telephone, to transmit information. We could use long, strong molecules like polyethylene or buckytubes for the strings. Crossing strings could actually touch without a problem, if designed right—waves traveling in different directions can go right through each other. The strings could easily be less than a nanometer wide.

Tne speed of sound in *diamond* is over ten miles per second; in appropriately stretched strings it might be even faster. If you design your mechanical nanocomputer in a conservative mode where vibrations are allowed to die out between operations, it could run with a gigahertz clock speed. An optimized design where all the wave propagation and reflection are taken into account might manage into the hundreds of gigahertz. Small, carefully tuned "circuits" could operate at a terahertz.

STORAGE

One form of computing that is tantalizingly almost possible today is using DNA. The DNA molecule is a memory device, not unlike a tape in the tape drives you see on the pictures of the big 1960s IBM computers. The DNA in a given bacterium holds about a megabyte, and that in a human cell holds about a gigabyte. (The DNA is the same stuff in each case—it's just that the human cell holds one thou sand times as much DNA as the bacterium.) For comparison, this book is about a megabyte if represented as plain ASCII text and corn pressed pictures. So bacteria come with an instruction manual and humans with a thousand-volume set.

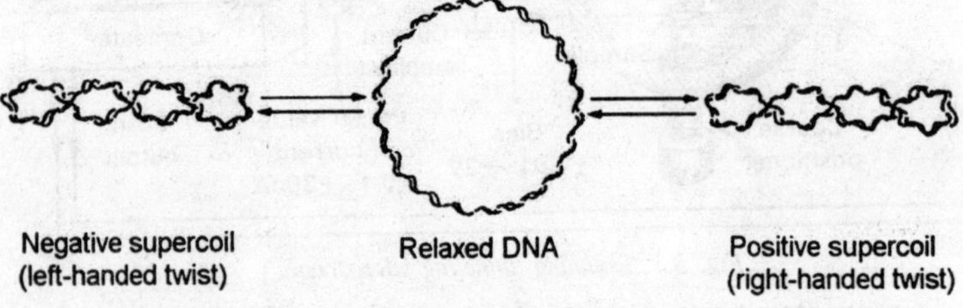

Fig. 3.2. A schematic diagram of a relaxed circular DNA.

The natural machinery in the cell has several mechanisms for treating the DNA as a *library*. It avoids the library's biggest headache, books being out when needed or not being returned, by copying off information that is needed (onto RNA) and distributing the copies. The DNA itself is copied when the cell divides. It is proofread and its errors corrected. There are mechanisms to roll the DNA up for compact storage and unroll it for reading. As part of this process, it's partially cut so one part can be twisted without twisting another, and then the cuts repaired.

Unless you've just had it sterilized, there is almost certainly more data storage in the DNA of the bacteria lying around on your computer, notebook, or PDA than inside as part of the device itself. At this writing, Google indexes about four billion Web pages. If we estimate the average size of a page, with pictures, as 125 kilobytes, the total comes to about half a petabyte (500 terabytes). If stored on DNA, the whole thing would fit in a 150-micron cube, the size of a grain of fine sand, or in other words, a speck just barely big enough to see with the naked eye.

Nanomachines will be able to store information more compactly than DNA, but not all that much more compactly. Using the same basic method, a series of different but similar molecules chained together, one can use smaller molecules than the *deoxyribose* ones in DNA, but there's not much headroom. One constraint is likely to be the competing desideratum of reading the information faster. That IBM tape drive was a lot bigger than the tape reel itself. The really dense part of the information storage system, the tape library, had racks of tapes nestled snugly together, but had a long access latency: some human had to walk down the aisle, pick a tape out, and mount it on a machine. *Nanocomputers* could have a similar scheme, with *nanorobots* retrieving tapes, and have latencies shorter than current-day hard disks.

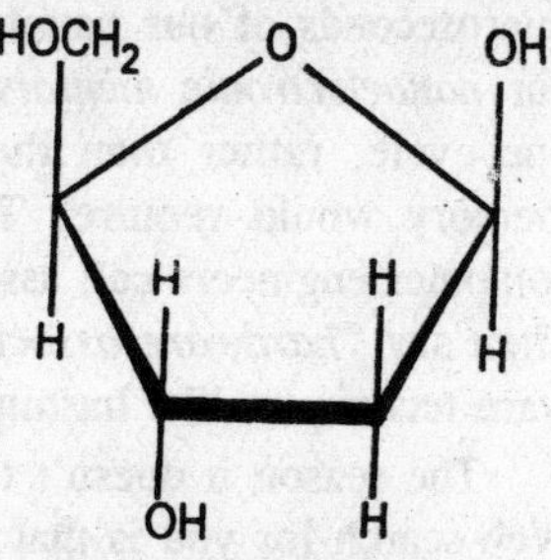

Fig. 3.3. Deoxyribose sugar.

As a general technological truism, memory gets bulkier as it gets faster. This will continue to be true at the nanoscale, but there's so much room at the bottom that this will not be a significant problem until expectations have risen quite a hit. The fastest memory in common use is the registers and cache that operate at processor clock speeds, currently in the low gigahertz range. These are made of the same logic circuitry used in the processor. For *nanoelectronics*, a bit using molecular switches, probably capable of operating well in excess of a terahertz, might occupy a 10-nanometer cube. This is one thousand times less dense than DNA, meaning that you'd need a grain of coarse sand to store the whole World Wide Web instead of a grain of fine sand. Sorry But you'd be able to access it in the amount of time it takes light to go a tenth of an inch—for the tape reel scheme, a beam of light could go twenty mile you wait for your data.

Neither of these times is perceptible to a human, but the difference shows up when you start adding them together, as in searching the entire database, collecting statistics, data mining, and so forth. Either form could read the database at a terabyte per second (tapes are much more efficient for sequential than random access). It'd take about ten minutes to get through the entire Web. For use as a server handing out Web pages to all corners, the electronic form could retrieve eight million pages a second. The tape form could manage only ten thousand. The amount of information the human brain stores can be estimated at something like 100 terabytes, within shouting distance of the whole World Wide Web. The brain has a cycle time closer to the 100 microseconds of our nanotape library system than the *picosecond* of our *nanoelectronic memory*; but it can search its entire contents in one cycle, rather than the ten minutes that either computer style memory would require. The brain uses something closer to what computer engineers call associative memory; (Software engineers will often say "*hardware associative*" memory to distinguish it from soft ware techniques like hashing, indices, and inverted files.)

The reason it doesn't take Google ten minutes, or more, to do a Web search for you is that it's prepared an index ahead of time. This is fine for something like the web, which changes slowly enough that indexing is a feasibly small amount of its work, and the kind of query you can ask is restricted to matching words. A hardware associative memory is one that acts as its own index, at the same speeds that ordinary memory simply retrieves words by their addresses. Associative memory is commonly used to maintain the indices of cache in *microprocessors*, where it would he too slow to use software techniques, and for Internet address tables in high-end routers, and in similar time-critical applications.

If the Web were stored in *nanoassociative memory*, you could ask things like "What's the average temperature that's referred to in Fahrenheit on pages where 'Fahrenheit' is misspelled?" and get the result back in nanoseconds.

Today, associative hardware is limited to the most demanding applications in computing, but it is ubiquitous in the nervous system of any animal, more complex than an ant. This is because nature uses nanotechnology to build her computers and can afford to throw lots of hardware at tough problems. Interpreting sensor data, in particular; is a hard computational problem. *Associative computers* are just plain smarter, cycle for cycle, than conventional ones. When we have nanotechnology to build our computers with, we'll have the luxury of throwing some extra hardware around and be able to build smarter computers as well as faster or more capacious ones.

Reversibility

Nanocomputers whether mechanical or electronic, will have to be designed differently from conventional ones at a logical level. The reason is *reversibility* and *thermodynamics*. Every time you erase a bit in any computer, you must dissipate some energy, the amount of which depends on the temperature. At room temperature, it's about four zettajoules. This lies right in the range of the energies that atoms have owing to their thermal motion.

So if you had a nanocomputer that had a thousand atoms per gate, and each gate erased one hit per cycle, and the computer had a 1-gigahertz clock, it would try to warm up at a rate of a million degrees per second. You could beat this by brute force if you had to: a small processor, with ten thousand gates, would dissipate only 40 nanowatts. You could simply cool it with a radiator. But cooling arrangements will complicate the design, push the computing elements apart to allow for pipes and so forth, and slow the whole system down.

It's best to avoid erasing those bits if you don't have to. In a simple, one-instruction-per-cycle computer, you don't logically have to erase more than about one word—typically thirty-two bits today—each cycle. All the others can be uncomputed rather than simply erased. This is the electronic (or mechanical) equivalent to compressing the bicycle pump until it's hot before putting it into the hot bucket, and so forth. If care is taken, most of the logical operations in the computer can be made reversible. By making the logic of most of the processor's instructions reversible as well, we should be able to design computers that erase only one or two bits per cycle instead of tens of thousands.

Writing programs in a completely reversible instruction set is a bit tricky (although quite possible). Writing with a mostly reversible construct set is not much harder than ordinary assembly-language programming. It's quite straightforward to write a compiler that turns your higher-level language code into mostly reversible machine code without you even knowing about it. So while reversibility will matter to machine designers and compiler writers for nanocomputers, ordinary programmers, much less people who just use the software, won't know or care.

Quantum Computers

The reason that *quantum mechanics* is so expensive to simulate is that quantum systems can act as if they are doing lots of different things at the same time. The classic example is that you can fire an electron at a plate with two slits on it, and it acts as if went through both of them and interfered with itself. This is called superposition of states. The basic idea of *quantum computing* is a simple extension: let build a computer that acts as if it were doing lots of different computations at the same time.

A fairly substantial research effort is attempting to do this today. It faces substantial challenges. The first is theoretical: although the quantum machine could do many calculations, you can get the results of only one of them. The electron, having gone through both slits, will hit only one spot on the screen on the other side. It'll always hit in a spot where it would have been possible to go both ways and not cancel itself out, but out of such possible spots, one is chosen seemingly at random.

Thus you can't use the *quantum computer* as a generally capable *multiprocessor* with as many processors as there are superpositions. But you can do computations that would have involved trying out a huge number of possibilities, when all you need is the one that worked. The classic example is searching for encryption keys, but many other problems can be posed in a similar way: finding the path through a maze that reaches the exit; finding the sequence of chess moves that leads to a checkmate; finding a machine design that per forms a given task; and so forth.

There are practical difficulties as well. To use quantum superposition to get the multiple computations, you have to be using single quanta, such as photons or electrons, to represent your bits. Thus you have to be able to manipulate them individually. In a macroscopic lab, it's difficult to handle more than a few of these in a way where unwanted influences from the rest of the universe won't slip in and mess up your computation (a process known as *decoherence*). This is where nanotechnology might help: building quantum computers with more superposed bits (called *qubits*) and where the coherence lasts long enough to perform a useful computation.

Quantum nanocomputers will blow the doors off any classical one for the kinds of problems they can do. Because of the limitations of quantum algorithms, however—you can do a vast number of computations simultaneously but get only one of the answers—they will remain special-purpose machines.

4

SELECTIVE APPLICATIONS

Some new technologies change every aspect of our lives. *Electronics* is probably the best known contemporary example of this sort of changes. Electronic machines have changed the way we work, play, and live—or don't live. They have created whole new industries, including the concept of a "*knowledge worker.*" The information industry, one of the hallmarks of modern civilization, could not have grown to its current size and complexity without electronics. There would be no computers without electronics—Charles Babbage notwithstanding. There would still be entertainment, of course, but there would be no radio or television or stereo.

Nanotechnology is a technology destined to change every aspect of our lives. It will allow marvelous new machines and applications, things we can only just imagine, and things we can't imagine at all. It will have an overwhelming impact on our daily lives. And it will happen sooner than you think. Fairly simple, easily engineered nanomachines are capable of performing tasks that have value—or at least impact—at virtually every level of our lives. We don't need to develop full-brown replicating assemblers to enter an entirely new world. It may be that what is described in this chapter is all that can be predicted. As others have pointed out, once self-enhancing *artificial intelligence* (AI) has been built, we quickly snowball into a sort of singularity. It is very difficult to predict what entities much more intelligent than you or I will create, or for what purposes they will use the things they do create. *Artificial intelligences* that can redesign and improve themselves will develop at exponential rates, and it is quite possible that such intelligences will appear shorty after nanotechnology creates molecular-scale computing devices, devices with truly awesome power.

This sort of AI is not only unpredictable in its capabilities. It's unpredictable in its intent. Our only hope for survival may well be to get there first by creating enhanced human beings with equally high levels of intelligence. But, by definition, such *intelligence amplification* (IA) will produce enhanced humans that are only partly human—and the human part will not be greatest part. Again it becomes difficult to predict what such entities will want from their technology. And it is hard to predict what they will do with nanotechnology.

This isn't meant to sound pessimistic. It will be very exciting. But one of the reasons for excitement is the fact that it's unpredictable. In the meantime—and it may take several decades—we'll have the technology, machines, and applications described in this chapter to keep us busy. It aims to cover the territory from the first realization of nanotechnology to the creation of true, inexpensive replicators. After that, it's wide open. You may find glimpses of the future here. Or you may find that we're in the same position as a prehuman ape trying to imagine what electrons are good for. We certainly don't have the perspective, and we may not even have the intelligence or imagination —yet.

NANOMACHINERY

There are usually several ways to engineer a device, *nanoscale* or *macroscale*. The final design is based on hundreds if not thousands of practicalities, efficiencies, availabilities, and experiences. Since many of these are currently unknown for nanotechnology, this presents a plausible rather than certain design for a general purpose nanomachine that can be programmed for several different uses. The following paragraphs present a description of the scale, shape, processing power, communication systems, energy requirements, and construction techniques of one such nanomachine.

Scale

Although the active elements of this device are measured in nanometers, the machine itself measures approximately a micron or a millionth of a meter. There are several reasons for this. The machine needs to be resistant to damage and capable of affecting its environment without breaking up. In addition, the device needs a reasonable amount of computing power to accomplish its tasks, and computational machinery, especially memory, requires a certain amount of space.

The device also needs to influence the outside world, physically pushing against things or reflecting visible wavelengths of light. Finally, the energy requirements of many applications will require power supplied from outside the device, either electrically or mechanically. Such an interface requires a certain minimum size.

Shape

Spheres are the most efficient shape, in terms of volume-to-surface-area, and the most efficient and stable spherical approximations are the various geodesic forms studied by Buckminster Fuller. A micron-scale geodesic structure built with carbon rod struts would form an extremely secure exoskeleton for housing nanomachinery. Whether such forms are based on the popular soccer ball structure of buckminster-fullerence or other geodesics is a matter of design. In any case, a series of structurally linked, concentric geodesics could easily form a secure shell for nanomechanisms, providing them with structures to push against.

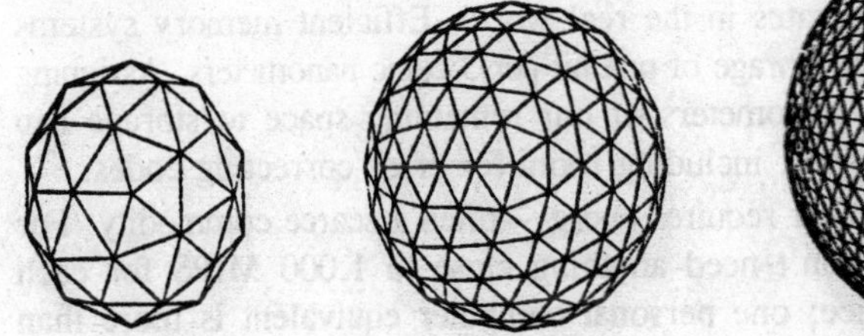
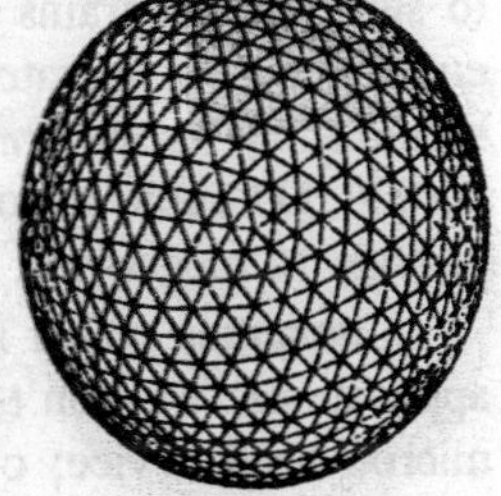

Fig. 4.1. Regular geodesic two-, four-, and nine-frequency icosahedrons.

A one-*micron sphere* provides a volume of about 520,000,000 cubic nanometers. Dedicating a third of that to structural supports—a conservative estimate—leaves just over 340,000,000 cubic nanometers for functional machinery.

By extending arms through the vertices of geodesic, the machine can influence and communicate with the outside world. Simple collapsible rods allow external objects to be pushed away and can also make electrical power contacts. Rods that end in protrusions—perhaps dynamically extensible themselves—allow other nanomachines to grab the rod. In this way, to nanomachines can both push and pull each other and establish secure power and communications linkages.

To give each rod a good reach, we can use a telescoping design. To allow them to fit snugly within the structural framework, each rod, when collapsed is a little over a third of the machine's total diameter. We then need nine telescoping components to give the rod a reach of three microns, or three times the diameter of the machine itself. If each rod component is a hollow cylinder with a wall thickness of ten nanometers, the entire assembly will fit in a cylinder 180 nanometers in diameter and 400 nanometers in length, and occupy about 10,000,000 cubic nanometers.

If the basic design is icosahedral, and one rod extends through each primary vertex, twelve rods are required, occupying a total of 12,000,000 cubic nanometers.

Processing Power

Each machine need storage and computing facilities. It seems that it will be possible to build a 1,000 MIPS (million instructions per second) molecular computer that fits inside a cube 0.4 microns on a side (i.e., in a volume of 65,000,000 cubic nanometers). This is roughly 1,000 times the computing power of today'' personal computers.

In addition to the processor, we need a large memory, primarily to store the programs that deal with the situations the machine may encounter as it operates in the real world. Efficient memory systems should allow for the storage of one bit per 5 cubic nanometers. Assigning 50,000,000 cubic nanometers of our remaining space to storage can provide 10 megabytes, including room for error correcting codes.

Computing power requires energy—often a scarce commodity. For applications, we don't need anything close to 1,000 MIPS for each micron-sized device; one personal computer equivalent is more than enough for our purposes. This reduces the energy needs a thousand fold. *Molecular memory* does not require large amounts of energy. Indeed, it only requires energy when it's being accessed. By reducing our requirements for computing speed, we also reduce our needs for space and energy.

Communication

These devices need to communicate with each other and with human users. In fact, both types of *communication* will be needed simultaneously as the machines interact with each other and the users to carry out the user's wishes.

For many applications, a number of *nanomachines* must work cooperatively toward a single goal. To do so requires communication between the devices. In some circumstances, it may also be desirable to employ large amounts of computing power. By cooperating, the machines can turn their millions of individual processors into a single, much more powerful, *multiprocessor computer*.

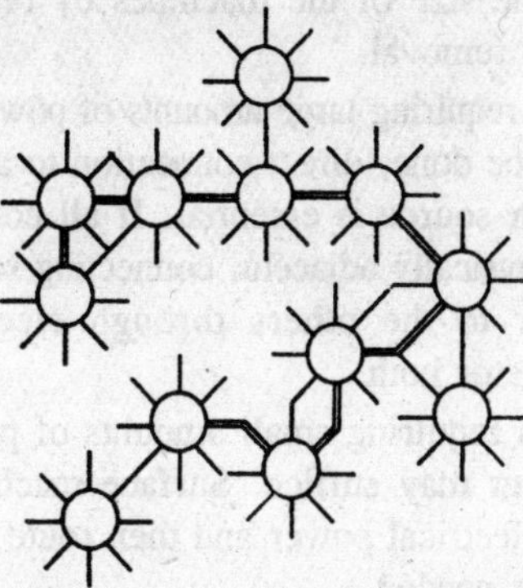

Fig. 4.2. Networking of nanomachines with communication paths shown.

Peer-to-peer communication is achieved most easily by electrical connectivity. Even a fairly sparse set of connections can provide global communication if the individual machines forward messages to other machines.

User communication is a different problem. In many applications—though not all—the user must tell the machinery what to do. This can be accomplished via *optical communications*, using the same technology that is used today for remote control of *stereo components*. Surface nanomachines detect the optical signals, interpret them, and pass them on to machines not in the line of sight of the control device.

Communications between machines is actually a computational issue. It uses the computing power described in the previous section and does not require any additional space. Communication channels can piggyback on structural and power components of the machinery so that no additional space is required either. However, we will need to allocate some 10,000,000 cubic nanometers per device for signal detection and amplification.

Energy

Some nanomachines can be designed to operate passively. But many will require power to actively affect the environment. In fact, this is probably the most limiting aspect of the nanomachines design problem, and one often overlooked when uses for nanomachines are considered. Although nanotechnology provides many solutions for mechanical problems, it provides few solutions for energy production—and nanomachines can be very energy thirsty.

In addition, heat dissipation can be a seriously limiting factor in the design of *nanomachines*. It can restrict the density of the machinery, or it can increase the size of the machines by requiring the addition of channels for heat removal.

For applications requiring large amounts of power, where substantial physical work must be done, direct connection to an external electrical or mechanical power source is essential. If all coherently functioning nanomachines are physically adjacent, connecting some will allow them to route the power to the others through electrical junctures or mechanical couplings or both.

For applications requiring small amounts of power, *photocells* on *surface nanomachines* may suffice. Surface machines would convert available light into electrical power and then route the electrical power to other machines as needed.

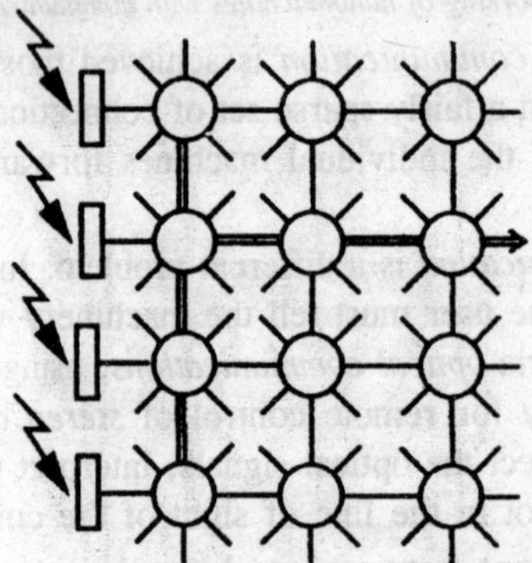

Fig. 4.3. Machines with photocells routing power to other, interior machines.

For applications requiring only occasional bursts of power, or operating in environments with a great deal of physical motion, simple physical shaking may be sufficient to supply the needed powers—like an old, mechanical, self-winding watch. In some cases, the user may literally shake the nanomachines before using them. In other cases—in an automobile or on clothing—normal environmental movements may supply enough energy.

Wherever the power comes from, and in whatever form, a mechanism is needed to store it and convert it to a usable form. About 15 percent of our nanomachine or 75,000,000 cubic nanometers, will be reserved for power storage and conversion.

Construction

As described, these nanomachines are very simple. They are composed of simple, nested geodesic skeletons and extensible rods, with no more computing power than today's personal computers. It is not such a tremendous task to design machines with *nanometer-scale components*, but how can we construct them?

At first, the answer will no doubt be, "*one at a time.*" As this writing, we can manipulate individual atoms at the nanometer scale and build more complex machinery at the micron scale (e.g., *computer chips*). Given the pace of technological development, we will certainly see machines of the complexity described here built "*by hand*" fairly soon. But to be useful, we will need trillions of them—one trillion would fill a single cubic centimeter.

In time, limited-scale mass production will arrive. Shortly after the first effective assembler has been built by hand, they will be manufactured in quantity—and used in turn to manufacture the machines described here. If a factory has ten million assemblers (which might have taken years to create), each capable of producing ten thousand

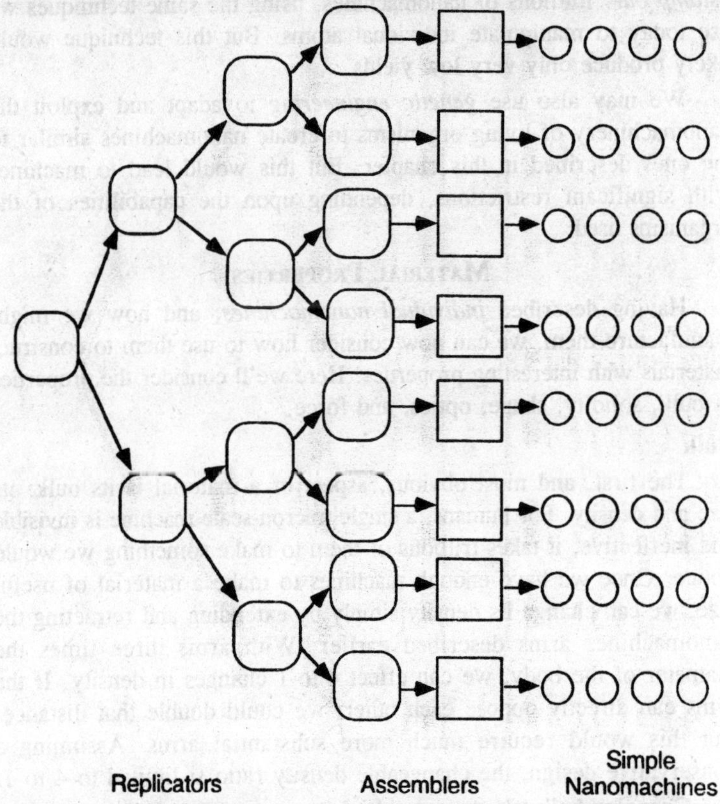

Fig. 4.4. Replicators producing assemblers producing nanomachines.

nanomachines a day, it will take two weeks to produce a cubic centimeter of finished goods. During this period, nanomachine-based products will be rare and very expensive.

Eventually, *replicators* will be used to produce assemblers, which will then produce nanomachines for final use. During this period, replicator technology will be viable only within very constrained environments. I expect that this period will last for some time, since designing unconstrained replicators is a very hard problem. Although this period will not allow consumers access to replicator technology, it will be a time filled with inexpensive, easily accessible nanomachinery of the type described in this chapter.

These three stages of nanomachine production seem most likely, but there are other possibilities. A crystal array might simultaneously "*stamp out*" millions of nanomachines, using the same techniques we use today to manipulate individual atoms. But this technique would likely produce only very low yields.

We may also use *genetic engineering* to adapt and exploit the nanomachinery of living organisms to create nanomachines similar to the ones described in this chapter. But this would lead to machines with significant restrictions, depending upon the capabilities of the organisms used.

Material Properties

Having described *individual nanomachines*, and how we might manufacture them, we can now consider how to use them to construct materials with interesting properties. Here we'll consider the properties of bulk, solidity, shape, optics, and force.

Bulk

The first, and most obvious, aspect of a material is its bulk: its size and density. For humans, a single micron-scale machine is invisible and ineffective, it takes trillions of them to make something we would notice. Once we have enough machines to make a material of useful size, we can change its density simply by extending and retracting the nanomachines arms described earlier. With arms three times the diameter of the body, we can effect 4-to-1 changes in density. If the arms can directly oppose each other, we could double that distance. But this would require much more substantial arms. Assuming a conservative design, the changeable density ratio si limited to 4 to 1.

Changing bulk takes energy. In the process of extending the arms, adjacent machines may need to be pushed away. If a good power source is immediately available, there's not problem, but power will often be in limited supply. Still, there are several possible solutions. By employing a ratchet mechanism, arms can be constrained to extend (or contract), so that density can be changed by shaking or "*fluffing*" the material. In situations where there is a lot of physical movement—a car's shock absorbers, for example—this is even easier. Or, if we're not in a hurry, low-level, external power sources such as sunlight may suffice.

Solidity

Viscosity is a measure of how much internal friction a liquid has, or how easily it flows, or, in a sense, how *liquid* the liquid is. We can adjust the viscosity of our machines by having them grab—or not grab-onto each other. In the extreme cases, we end up with a solid, where all the machines are grabbing each other firmly, or a fluid in which all the machines tumble past each other without constraint. Between these two extremes, we can set the granularity of the material and independently set of its resistance to flow.

In addition, we can also determine the strength of a solid. If all the machines are initially set to grab each other, they can be programmed to release as certain levels and types of forces are applied. We can make a very hard solid, or a crumbly one, or one that's very resistant to horizontal force but not at all resistant to vertical force. We could also make solids that are extremely solid at low pressures, but which give way at slightly higher pressures. Or the other way around.

Shape

Shape has two components: overall form and surface detail. Object form is the shape of the object itself—a cube or sphere or lawn chair or teapot. Surface detail determines the texture of the surface, the smoothness—or sharpness—or the edges, and so forth.

Nanomachines can coordinate with each of the build a particular shape, or, with much less expenditure of energy, they can let users set the shape. For example, a special tool could be used to indicate a plane along which to break. All machines along that plane would simply let go to their neighbours, so that the material breaks apart along a perfectly straight cut. The process can also be reversed to seamlessly recombine two pieces of material. Using techniques like this, arbitrarily complex designs can be fashioned without requiring the material to consume large amounts of energy or to solve complex global communication and coordination problems.

Textured, forming and maintaining desired edges and otherwise coordinating the surface structure requires an awareness among the nanomachines of the existence of the surface. Those machines at the surface an sense that fact because they are only in contact with other machines on one side. By communicating with other adjacent machines, they can determine the current shape of the surface and what needs to be done to reach the desired shape. As always, it may require some energy input to effect the change. A *nanoengineered knife*, for example, may need to be tapped against the counter a few times to supply the energy required to resharpen it.

Optical Properties

There are two main aspects to optics: changing the machine-level characteristics of a material and determining what this looks like to a human.

Micron-scale machines are large enough to reflect visible light. This is both a blessing a curse. On the one hand, it's fairly easy to change their reflectivity characteristics; on the other, transparency becomes a challenge. While complete transparency seems quite hard, a variety of optical effects could be derived by mixing the machines with inert, transparent tiles that the nanomachines could rearrange—covering them or causing them to overlap.

With this control, we can select an object's overall appearance. The easiest effect to achieve is changes in colour. Simply by telling the machines to randomly pick one of three colours, with probabilities supplied by the user, any colour is possible. With global coordination, the machines could display any image the user desires, in full colour, with micron resolution.

Force

Materials use *force* to change shape, in some cases to maintain shape, to absorb acceleration and deceleration, to vibrate, to move external objects, and to move themselves. Since often-expensive energy is needed to generate a force, we'll consider three situations: those requiring little to no energy, moderate amounts, and large amounts of energy.

Applying a force always requires energy form some source. So when we say we need little or no energy, we really mean we're being efficient about recycling energy that came in from outside, or finding ways to replenish the energy supply over time. For example, nanomachines can act as springs to absorb and then return energy—but not necessarily immediately, as a normal spring does. This can lead to materials that deform when pressure is applied, but return to their original shape when the pressure is relieved, either immediately, over time, or on command. With moderate energy, nanomaterials can significantly reshape themselves, change their texture or optics, or provide vibrational or other dynamic stimulation. The energy required might come from outside the material or from the user. For example, if you shake a "book" in a certain way, it could display different content. If energy is abundant, almost anything is possible. The material can reshape itself at will, carrying with it the user, other objects, or for that matter, the entire house.

5

ATOMIC CONTROL

A remarkable characteristic of molecular manufacturing is that it brings the problem of manufacturing together with the problem of materials. In classical materials science, one treats materials as a continuum of "*stuff*," and the manufacturing process then shapes that continuum of stuff into parts. In molecular manufacturing, the parts are the fundamental building blocks of matter, and so, making the material and making the component is all one process of joining these components to form solid structures. Mike Kelly is an entrepreneur and consulting professor at Stanford University in the Department of Materials Science and Engineering, a department in which working with atoms as components is of great importance, in which increasingly the importance of getting the atoms in the right place is being recognized, and in which the utility of that task is being exploited.

The impressive pictures that Ralph Merkle presented of computer simulations of bearing structures built of diamond or diamond-like materials raise the question of whether we could make something like that today with some kind of extension of current technology.

To address that question, it will be better to consider it in two parts. First, what are some examples of current systems that exhibit, to a greater or lesser degree, atomic control over fabrication in one dimension? By "one dimension," mean the thickness dimension of thin films. Second, how might these capabilities be extended to the fabrication of nanoscale devices?

Synthesis techniques, in general, can be categorized as shown in Table 5.1. The *classical kind* of materials science approach (which has developed almost everything that we have, from a slide projector to the Brooklyn Bridge) involves fabricating materials through some bulk process, and then forming or deforming that bulk material so that it can serve some function. The technologies of metallurgy and ceramics have played pivotal roles.

Table 5.1. General categories of synthesis techniques

	Classical material science	*Thin film*	*molecular manufacturing*
Approach	Bulk processes	One-dimensional atomic control	Three-dimensional atomic control
How to synthesize	Form bulk, then deform for function	Atomic layers, then each for function	Synthesize for function
Technical base	Metallurgy, ceramics	Solid-state physics, chemistry	Organic chemistry, catalysis, biology

More recently, thin-film structures have appeared, which are built an atom at a time in one dimension, and then etched or processed by some more crude technique to fabricate a functional device. The scientific disciplines that provide the basis for those techniques are solid-state physics and chemistry.

Eric Drexler's ideas on molecular manufacturing involve a third approach in which we will be able to directly synthesize (to atomic specification) three-dimensional structures designed to perform specific functions, without a need for further processing. Such molecular manufacturing is already done by biological systems, and certainly one promising approach to synthesizing molecular machinery is to follow the path shown by molecular biology.

Thin-film technology is a bottom-up synthesis approach (that is, materials are built one atom at a time). The methods normally used are vapour deposition techniques of one sort or another, specifically, physical evaporation and chemical vapour deposition (CVD) techniques.

Thin-film Technology

These techniques all involve a substrate that is placed in a vacuum and bombarded by particles. In the case of physical vapour deposition, a sputtering source or a thermal evaporation source is used to produce atoms or ions, which are by themselves very reactive species. These condense on a substrate to form a film. In contrast to the atomic species involved in these physical methods, CVD and its derivatives involve stable molecules that are brought to react on a surface. These

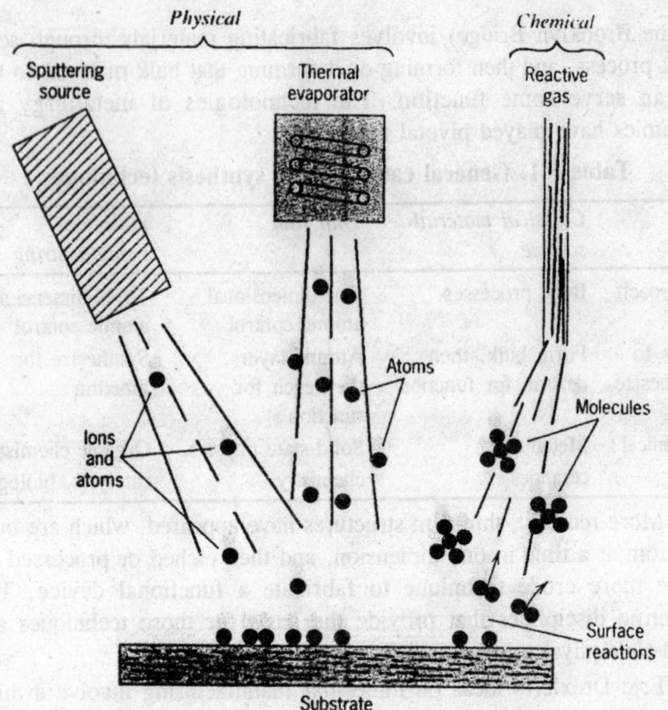

Fig. 5.1. Vapour deposition bottom-up synthesis techniques.

surface reactions leave a desired residue on the surface, and the other species that are formed then leave the surface.

Vapour deposition techniques have been used for quite some time to develop useful structures. For example, a multilayer optical film developed during the 1950s consists of alternating layers of transparent materials (in this case, zinc sulfide and magnesium fluoride) with different indices of refraction. Each layer is about 3,000 atoms thick.

Thin film has the property of being very reflective to visible light, but it transmits most of the light in the infrared and ultraviolet regions of the spectrum. Such a coating would be useful, for example, as a cool minor, and dental mirrors that are used to focus a lot of light in your mouth, with very little heat, might use such a structure. What stands out from a fractured, edge-on view of this particular film is that the quality of these early films is really very poor The surfaces are extremely rough on an atomic scale, and the crystalline structure of each layer shows many imperfections. While far from atomic control, these films nonetheless found use in a number of practical applications.

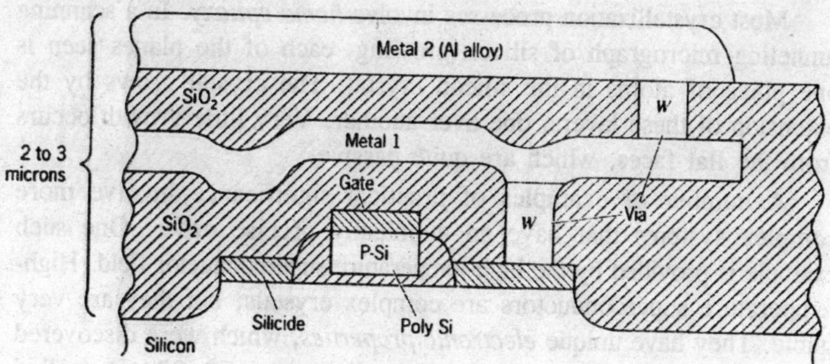

Fig. 5.2. Cross-sectional view of a field effect transistor (FET).

Probably the most significant use of this technology to date has been the development of integrated circuits. Fig. 5.2 depicts a transistor that is part of an integrated circuit. The structure is, in total, a few thousand atoms thick. It involves perhaps 10 layers of different materials. Some layers are metals, some are insulators, and some are very carefully controlled crystalline, doped semi conductors. Such a device is again synthesized an atom at a time in one dimension. However, to form structure in the x-y plane, it is necessary to revert to older techniques of lithography and etching.

Eric Drexler has spoken of the need for "hands" that can control things precisely on an atomic level. Such structures as integrated circuits might be said to be built by very clumsy, very shaky "hands." That is about the best that we can do today in fabricating three-dimensional structures.

Epitaxial Growth

How do we control the quality of the layers of an integrated circuit? Some of the layers in an integrated circuit are of quite good quality on an atomic level, while others are not so good. What we can often do is to use the substrate to provide intelligence to the deposition procedure. The regular pattern of atoms of the crystalline structure of the substrate acts as a template to impose order upon the atoms deposited on top of it. Quite precise structures can be made in this way. It is possible to use these deposition techniques to grow *amorphous films* (in which there is little long-range order in the deposited films) or to grow *polycrystalline films* (in which there is local order only) because on a larger scale, the film spontaneously divides into tiny crystals. However, it is also possible to grow films that are atomically perfect, and this third type of growth is called *epitaxial growth.*

Most crystallization processes involve *homo epitaxy*. In a scanning tunneling micrograph of silicon growing, each of the planes seen is one plane of atoms in the silicon crystal. The crystal grows by the extension of these layers, one over another. Very little growth occurs from the flat faces, which are quite passive.

A number of examples of practical significance involve more complex systems that have an atomically precise order. One such example is provided by the high-temperature superconductor field. High-temperature superconductors are complex crystals, but they are very stable. They have unique *electronic properties*, which were discovered by accident. The unit cell of one particular *superconductor* (called *YBCO*) that has a critical temperature somewhat above the boiling point of liquid nitrogen, shows a complex structure, with the unit cell containing about 50 atoms of 4 different kinds. The material must be precisely in this structure to be superconducting.

Among a number of ways of synthesizing these materials, one of the easiest is conventional "*shake-and-bake*" chemistry. If the proper stoichiometry of raw materials is ground up and heated, the crystal structure necessary for superconductivity results. It is not believed that the high-temperature superconductors would have been discovered if these crystals had been more difficult to fabricate than that.

It is also possible to grow quite good epitaxial layers of these superconductors. A *transmission-electron micrograph* showing three unit cells of a high-temperature superconductor on top of a magnesium oxide substrate. The atomic precision of the superconducting film and the alignment of the film with the substrate are quite good.

Solid-state lasers provide another example of the application of epitaxial growth to achieve atomically precise structure in one dimension. The sandwich of different materials that is used to make

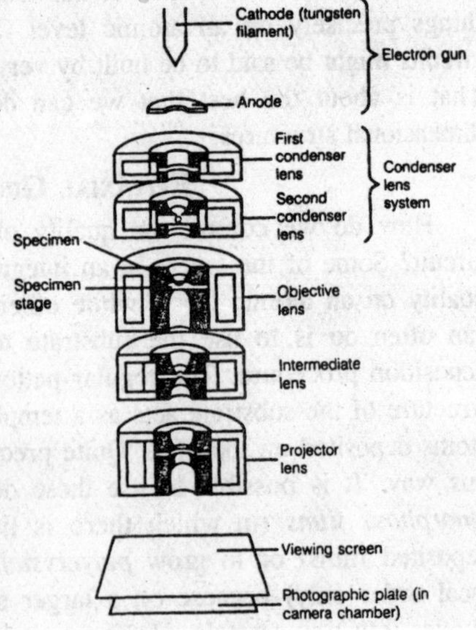

Fig. 5.3. Sematic diagram of transmission electron microscope.

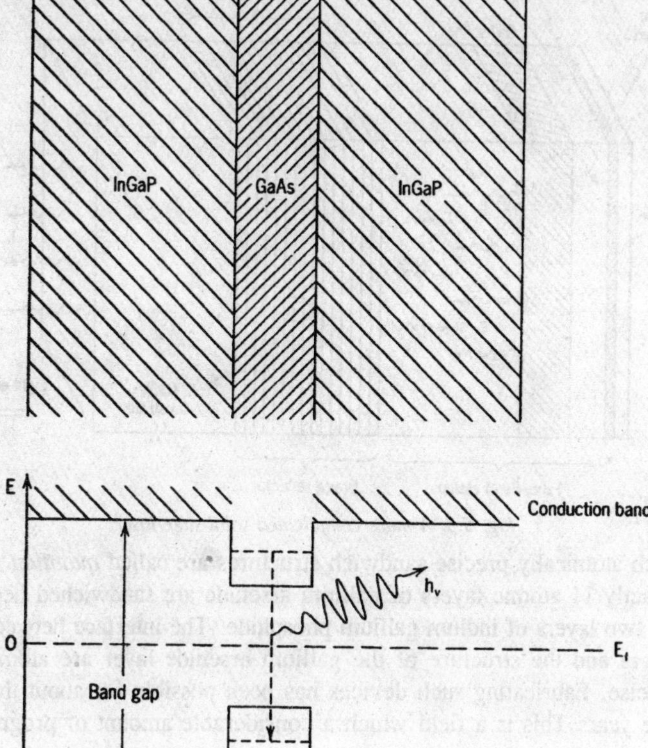

Fig. 5.4. Solid-state lasers.

a laser, such as the hand-held scanners that are used in supermarkets to read bar codes. Solid-state lasers use semiconductors that have two different band gaps (the band gap of the indium gallium phosphide is shown on either side, and the band gap of gallium arsenide is in the middle). The laws of quantum mechanics require electrons that exist in such thin layers to exist only is very discrete energy levels. A transition of an electron from a higher to a lower energy level occurs with the emission of a photon, providing the basis for the development of the laser The interesting feature of this device is that the wavelength of the laser can be varied by varying the thickness of the gallium arsenide layer.

To make solid-state lasers work well, it is necessary to have material i which the precision of the atomic layers is quite good.

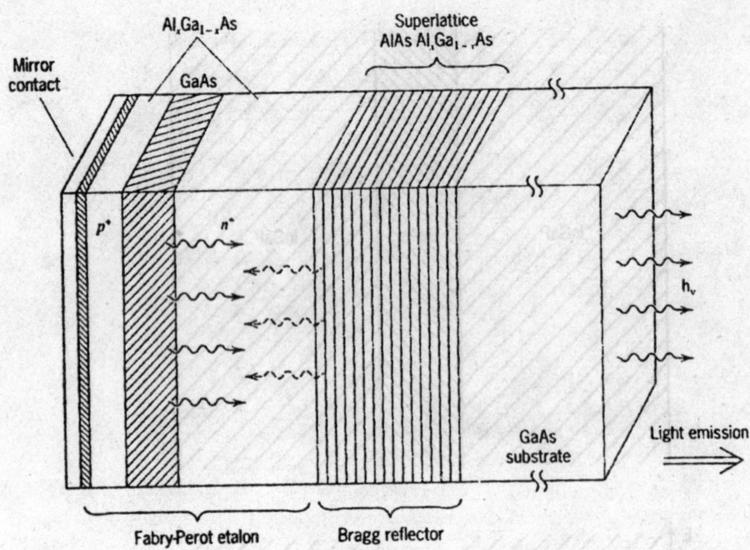

Fig. 5.5. A more complicated solid state laser.

Such atomically precise sandwich structures are called *quantum wells*. Exactly 11 atomic layers of gallium arsenide are sandwiched between the two layers of indium gallium phosphide. The interface between the layers and the structure of the gallium arsenide layer are atomically precise. Fabricating such devices has been possible for about the last five years This is a field which a considerable amount of progress is being made.

A somewhat more complicated laser structure is shown in Fig. 5.5. A gallium arsenide quantum well, similar to the one shown in Fig. 5.5, is shown the left of the structure. However, this particular device incorporates a mirror to the left, and a stack of optically refractive layers to the right of the quantum well. These layers are a partial reflector, which lets out a small portion of the light, but which forms an optical cavity to keep the lasing action going.

Atomic Layer Epitaxy

Another technique gaining some prominence in terms of improving the degree to which deposited films are atomically precise is called *atomic layer epitaxy* (ALE). A very good flat-screen display currently being marketed uses zinc sulfide as a luminescent material. The zinc sulfide must be a very good quality of material in order to produce light efficiently and without dead spots. Conventional deposition of zinc sulfide would involve two evaporation or sputtering sources that bombard a surface with zinc and sulfur atoms. This would grow a surface of about a thousand atomic layers to form a zinc sulfide film that would spontaneously crystallize.

ALE makes it is possible to do a much more precise job of making a zinc sulfide film. The essential idea is to put the components down sequentially. First, a large excess of zinc is deposited on the surface. If the bond formed between the zinc and the substrate is stronger than the bond between zinc atoms, then some temperature exists at which the upper layers all will evaporate, leaving a very clean and very complete single atomic layer of zinc. Next, an excess of sulfur is deposited upon the zinc surface. If the bond between zinc and sulfur is stronger than the bond between sulfur atoms, then again some temperature exists at which the upper layers of sulfur will evaporate, leaving a single layer of sulfur on top of the single layer of zinc. This process can be repeated to grow materials that have very good atomic purity and a very precisely layered structure.

ALE has several important advantages. First, because each added layer is first supersaturated, the layer that results has very few defects. Second, the use of chemical manipulations makes possible the growth of a large range of films. Third, it is possible to get very uniform coverage, even over convoluted surfaces. For example, to plate a film inside a trench formed in the substrate, conventional deposition techniques are not effective in depositing atoms on the inside of the trench. Most of the atoms will be deposited at the top, and not many atoms will diffuse to the bottom of the trench. However, with ALE, each added layer can be supersaturated, and the excess can be evaporated off, so that each time one atomic layer of reactant is deposited over the entire surface of interest.

ALE holds considerable promise for the future in developing atomically correct, layered structures. At Stanford, we are currently trying to grow diamond films using a variant of ALE. We are trying to tailor a sequence of reactions that, when completed, will give one layer of diamond. The goal is to grow uniform, ordered layers of diamond, grown perhaps at lower temperatures, and perhaps faster than with current techniques, but certainly with fewer defects.

A sequential vapour deposition reactor that we are using for this particular study. The substrate plate is tilted to the back and four ringed regions can be seen in the foreground. The substrate (which is usually silicon in our work) is placed in the hole in the substrate plate, and, as the plate is lowered over the 4 circular structures, the substrate is made to rotate rapidly (at 200 rpm) over each of the 4 regions defined by the 4 circles. Two of these are blanks and have not yet been used. The otber two are emitters—one a carbon sputtering source and the other a hydrogen source.

As Mike Pinneo describes that diamond growth involves the use of excited hydrogen gas and some form of carbon. We have been successful in growing films in which we have deposited about 1 monolayer of carbon on a surface, and then exposed the surface to atomic hydrogen (hydrogen gas which has been heated to about 2,500°K to dissociate it into neutral H atoms) in another one of these 4 emitter structures. This exposure converts the deposited carbon into diamond by some mechanism that is not fully understood. The cycle is then repeated, depositing more carbon and then converting it into diamond.

Using this technique, we have been able to grow good quality diamonds, by the 1,331 cm^{-1} Raman peak. Unfortunately, however, we have not bee able to get the process started correctly. We clearly do not have a uniform layer of diamond deposited on silicon. If we had a way to get that first layer of diamond to grow on the silicon substrate, then I think that this procedure would allow a homo-epitaxial growth of diamond that would be very uniform an would produce good quality films, which would have many practical applications. As yet we are not able to do that, however.

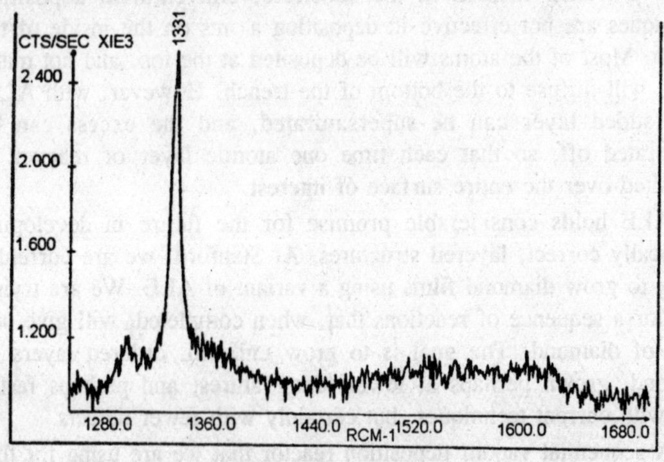

Fig. 5.6. Diamond grown by ALE.

Future Potential of Improved Multilayer Structures

A great future lies ahead for these precise, one-dimensional monolayers. One can imagine capacitor energy-storage devices that would rival batten made from atomic-scale inter layers of dielectrics and conductors. Other potential uses include faster and more dense semiconductors, and also better information storage devices, in particular using *magnetic thin films* only a few atom layers thick.

However, this approach still falls far short of achieving a general molecular manufacturing capability. Let us consider, as an intellectual exercise how we might use current techniques to try to grow one of the diamondoid bearings that Ralph Merkle described. Let us assume that we have perfected lithography to the point where we can make resists that are one atomic layer thick (for example, by using fluorine or hydrogen to passivate a silicon surface so that very little sticks to it). Let us further assume that we take advantage of smart substrates, in that we use the order existing in the material on which we are growing to impose molecular order on the layer that is being deposited. Lastly, let us assume that we have one steady hand—that is, that we have an STM that has reproducible atomic precision, and that allows us to place atoms where we want to place them, one at a time.

Could we use such tools to produce a diamondoid bearing? The first step is to use the STM to draw a pattern, such as the outside

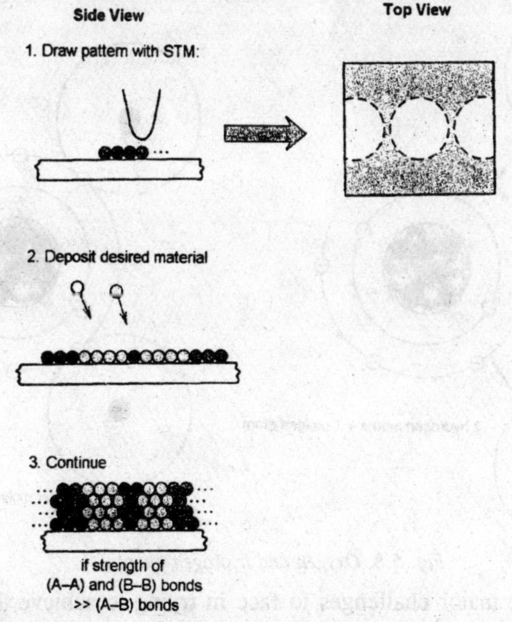

Fig. 5.7. Speculative fabrication of a diamondoid bearing using monolayer resists and ALF.

race of the bearing. We place one atomic layer of resist material everywhere we do not want the bearing to grow. For the moment, we will ignore the fact that oxygen and hydrogen atoms are also needed in Merkle's design. We will only try to grow the diamond core. The next step is therefore, to deposit carbon in the areas not covered by resist. Now we run into a really sticky problem that requires a little magic to solve. How do we make the carbon turn into not just diamond, but a highly stressed type of diamond with well-defined dislocations in exactly the right places within the diamond sheet? Moreover, we must face the question of whether one sheet of diamond is stable. Ralph Merkle's calculations show that the whole bearing is stable, but that result does not say anything about a section of the bearing.

If we assume that, through some magic, we could solve those problems, we could then continue to put more resist or scaffolding material in successive layers on top of the first resist layer, and continue to grow carbon layers within the resist to give the basic carbon structure of the bearing. Then we would have to liberate the bearing by selectively removing the resists and removing the substrate. Finally, we would have to add the required hydrogen and oxygen atoms.

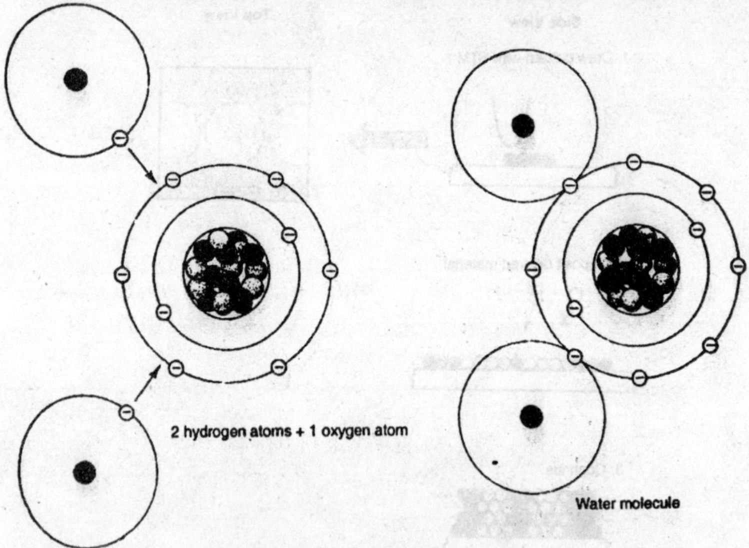

Fig. 5.8. Oxygen and hydrogen atoms.

We have major challenges to face in trying to achieve the three-dimensional fabrication of individual objects using current techniques. One challenge is how to identify enough atoms sufficiently quickly by using the STM (or AFM) to make such a fabrication practical. Two more significant obstacles require real breakthroughs for such a scheme to work. First, we must learn how to control the structure of the deposited material, which might require learning how to stabilize one layer of diamond. Second, we must learn how to incorporate other atoms into the structure (such as oxygen and hydrogen).

If generalized manufacturing of three-dimensional objects appears to be too difficult with these planar techniques, then what can we do with "one steady hand" that would be practical and important? Perhaps we could lay down a chain of atoms between two conductors, on some insulating substrate. The solid-state physics involved in describing the band structure of these linear atoms is quite different from what one thinks of in terms of bulk materials. Nevertheless, it is reasonable to imagine that, by choosing the right atoms and the right substrate, it would be possible to get a structure in which there are unoccupied states through which electrons could easily travel down this path. Thus, if one injected an electron at one end of the string of atoms, it would travel through the string of atoms to the other conductor. Such a line of atoms would be a semiconductor. On the other hand, if the state were partially filled, then the line of atoms would be a wire.

Accordingly, it should be possible in this way to make very small pieces of semiconductor and very small pieces of wire. The transit time of an electron through such a line of atoms would be on the order of 10 femtoseconds (1 fs = 10^{-15} seconds), so such a device would be very fast. The question then arises, how small is it possible to make a transistor if it is possible to make a wire or a semiconductor with just a few atoms?

The horizontal line of atoms is a *semiconductor*, as before. The shorter, vertical line of atoms is another semiconductor—or better, a conductor, that approaches but does not touch the first line of atoms. A small bias (perhaps a volt or half a volt) placed between the conductor on the top and the conductor on the right will cause a negative charge to propagate down the vertical wire.

A consideration of the effect of this charge upon the probability of *electrons tunneling* along the path from left to right shows that this structure would be a field-effect transistor that involves only three or four atoms. Thus, it should be possible to build extremely small semiconductor devices. The addition of another vertical wire approaching the horizontal semiconductor from the bottom would give a NOR gate. Adding some box that can contain or not contain an electron gives a memory element.

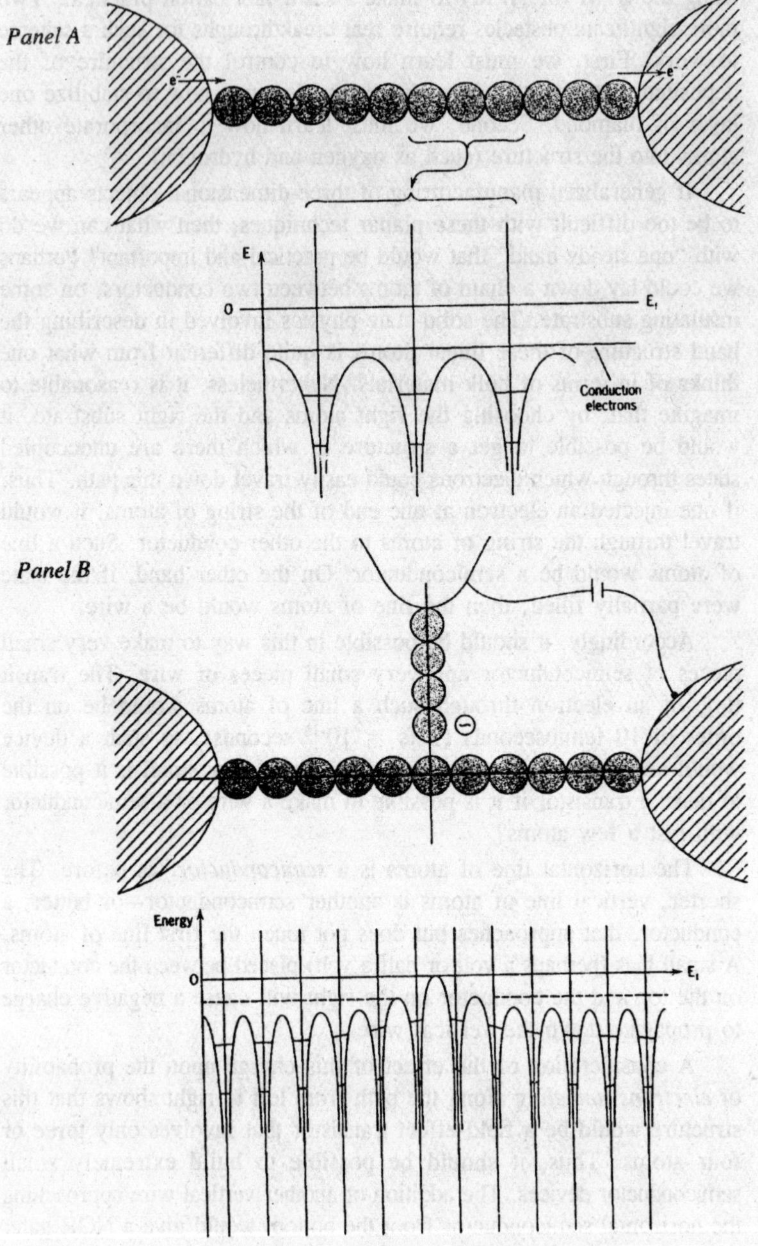

Fig. 5.9. The ultimate electronics.

Consequently, atomically precise control in only one dimension, with feasible improvements in current technology, could yield some real benefits, and could perhaps provide pathway technology along the way to more sophisticated nanotechnology.

While not involving lines of single atoms precisely arranged, as suggested here, interest is growing rapidly in fabricating and studying very small linear structures with widths on the order of a nanometer or so ("*quantum wires*" and "*quantum dots*"). These structures exhibit quantum phenomena (as do the quantum well lasers described previously) and suggest that useful devices can be constructed even without the atomic precision suggested here.

SUMMARY

In summary, the tailor surface reactions now to produce near-perfect, one-dimensional composites. The techniques that we are developing promise to be of use in fabricating three-dimensional structures if we can develop good, steady hands for working at the atomic scale. However, to extend this technology to build the molecular-scale machinery that Eric Drexler has described will require some key innovations. Perhaps, through the efforts of interdisciplinary groups striving to implement sophisticated nanotechnology methods, such innovations will appear.

6

POLICING SCIENCE

The paper opens with the question raised by Grundmann and Stehr, as to whether "*knowledge policy*" may include "the aim of limiting, directing into certain paths, or forbidding the application and further development of knowledge". It then explores this theme with reference to contemporary developments in biotechnology and nanotechnology, where the objective of knowledge is to enable us to create and modify at will biological entities (including humans and combined species known as "*chimeras*"), as well as self-assembling mechanical entities, ab initio through recombinant DNA techniques. I argue that a new category of risks is created by the promised technological applications of these forms of knowledge, called "*moral risks*", which threatens the ethical basis of human civilization; these are also "*catastrophic risks*", in that their negative and evil aspects are virtually unlimited. The paper asks whether our institutional structures, including international conventions, are robust enough to be able to contain such risks within acceptable limits; or alternatively whether these risks themselves should be regarded as unacceptable, a position which would impel us to seek to forbid individuals and nations from acquiring and disseminating the knowledge upon which those technologies are based.

A new *political field* which poses "the question of social surveillance and regulation of knowledge". They suggest that "*knowledge policy*" may include "the aim of limiting, directing into certain paths, or forbidding the application and further development of knowledge". If scientific knowledge is included here, it is, this proposition will not be well received. One of the great founding faiths of modern society is that of the infinite benefits of the liberation of the natural sciences from the intellectual and institutional shackles of dogma, including religion; its inspirational image is that of Galileo before the Inquisition, forced to recant publicly his belief about earth's movement in space, but unyielding in his mind and certain subjectively of his ultimate vindication. Anyone who seeks to challenge this faith is in for a rough ride.

Are there forms of knowledge about nature (including a technological capacity to manipulate nature based on them), now envisioned as practical possibilities in foreseeable futures, of which it may be said that they are too dangerous for humanity to possess? Too dangerous, at least, in the hands of that radically imperfect humanity in and around us, including its all-too-delicate veneer of civilization, which now seems prepared to seek that knowledge? And if so, is it even conceivable that one could argue for their suppression on the grounds that, once realized they will inevitably be deployed, to ends so evil, running unhindered into the future, as to destroy the moral basis of civilization? I at least am not ready to answer these questions—although they are being raised by some in the academic community, especially with reference to biotechnology. An editorial earlier this year in *New Scientist*, commenting on the inadvertent laboratory creation of a virulent engineered virus which could be used as a weapon in biological warfare, said:

There's also the problem that many biologists choose to ignore biotechnology's threats.... John Steinbruner of the University of Maryland, College Park, has suggested setting up bodies to oversee areas of biological research. Such bodies could question or even stop research, or decide if results should be published. As Steinbruner is well aware, his proposal strikes at the heart of scientific openness and freedom. But leaving things as they are is not an option. Biotechnology is beginning to show an evil grin. Unless we wipe that smile from its face, we'll live to regret it.

Here on specific themes with reference to a number of potentially catastrophic risks—risks having a dimension that calls into question the future of humanity itself—related to advances in contemporary scientific knowledge.

Catastrophic risk can be defined as the possibility of harms to humans and other entities that call into question the future viability of existing animal species, including our own. Thus these are not only risks to the present generations of living animal species, but also to future generations of presently existing species. One well-known risk of this type is what has been called "*nuclear winter*", the threat of a pervasive environmental catastrophe that could follow a large-scale exchange of nuclear weapons between the United States and the former Soviet Union (now Russia), under the doctrine of "*mutually assured destruction.*" The hypothesis of environmental catastrophe was based on the expectation that the earth's atmosphere would become loaded with particulate matter, blocking much of the solar radiation reaching the earth's surface, perhaps for a period of years (such an event is thought to have occurred following the impact of massive asteroids colliding with the earth). In addition, of course, the huge doses of radiation emitted by these exploding weapons would have profound genetic consequences for plants and animals.

LORDS OF CREATION

Given the existing stockpiles of nuclear weapons, the risks associated with them still exist, although it is difficult to know whether the probability now is greater or less than before. But new *catastrophic risks* are on the horizon, and these have a fundamentally different character that may require very different institutional responses from us. Their common characteristic, considered as basic and applied science and the technological applications made possible through them, is that they are all based on our latest understanding of biological systems through molecular biology. More specifically, their common scientific basis is the capacity to characterize complete genomes and to manipulate them by means of *recombinant DNA techniques* (or to create DNA-like mechanical structures).

The ultimate goal, already envisioned and set as an objective for research, is a knowledge of *genomics* so complete that living entities (and life-like mechanical entities) could be constructed, or alternatively deconstructed and then rebuilt and varied, *ab initio*. According to an article, researchers working with a microbial parasite sought to characterize and develop "an organism with a minimal genome, the smallest set of genes that confers survival and reproduction":

But since each of the 300 genes found to be essential could have multiple functions (*pleiotropism*), investigators had no way of finding the degree of redundancy and whittling the genome down further. The next logical step: make a synthetic chromosome of just those genes to build a living cell from the ground up.

Considered in their human implications, "*Moral risks*" are the catastrophic risks. Gradations of being (inorganic and organic matter, plants, insects, animals, humans) are and always have been a foundation-

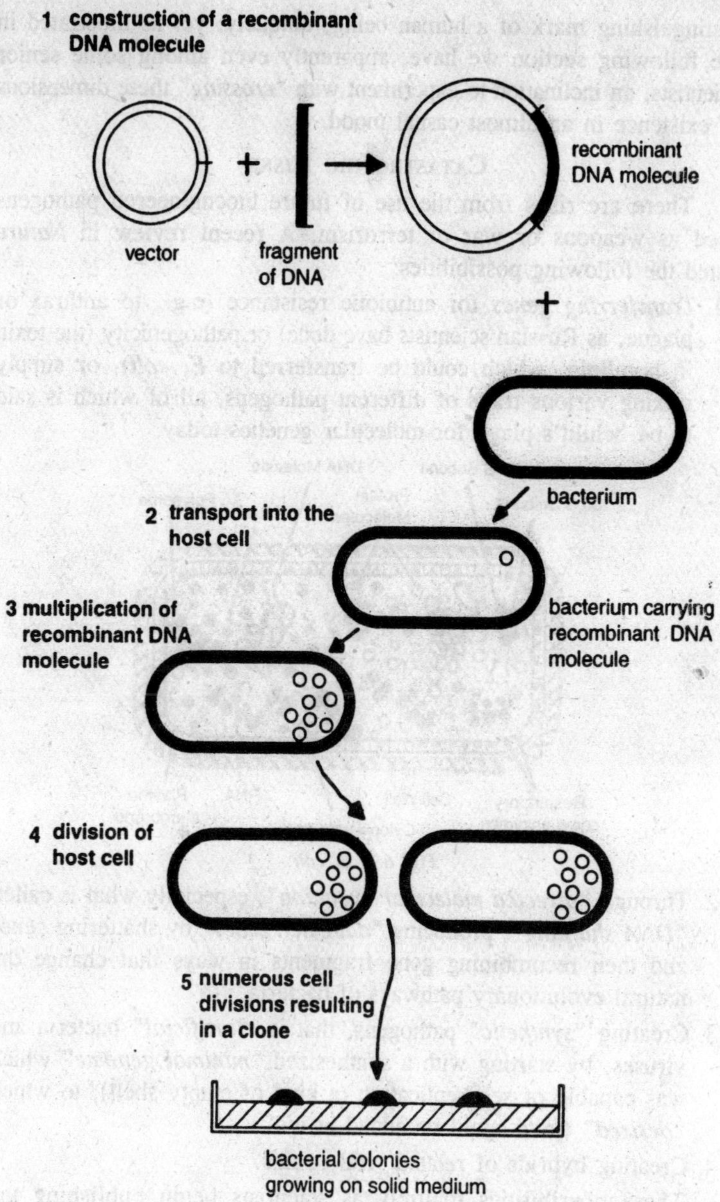

1 construction of a recombinant DNA molecule
vector
+
fragment of DNA
recombinant DNA molecule
+
bacterium
2 transport into the host cell
bacterium carrying recombinant DNA molecule
3 multiplication of recombinant DNA molecule
4 division of host cell
5 numerous cell divisions resulting in a clone
bacterial colonies growing on solid medium

Fig. 6.1. Recombinant DNA technology.

stone of humanity's ethical and religious systems. More particularly, "*self-consciousness*" has been regarded as the essential and

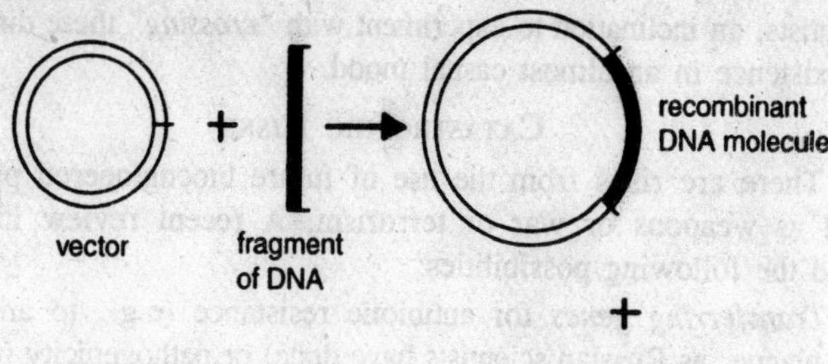

distinguishing mark of a human being, uniquely; yet as illustrated in the following section we have, apparently even among some senior scientists, an inclination to experiment with "*crossing*" these dimensions of existence in an almost casual mood.

Catastrophic Risks

There are risks from the use of future bioengineered pathogens used as weapons or war or terrorism. A recent review in *Nature* listed the following possibilities:

1. *Transferring genes* for antibiotic resistance (e.g., to anthrax or plague, as Russian scientists have done) or pathogenicity (the toxin in botulinin, which could be transferred to *E. coli*), or simply mixing various traits of different pathogens, all of which is said to be "child's play" for molecular genetics today.

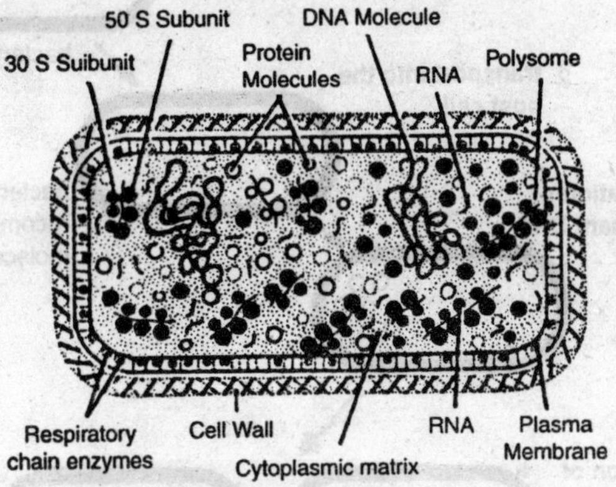

Fig. 6.2. E. coli.

2. Through "*directed molecular evolution*", especially what is called "*DNA shuffling*", producing "*daughter genes*" by shattering genes and then recombining gene fragments in ways that change the natural evolutionary pathways of bacteria.
3. Creating "*synthetic*" pathogens, that is, "*artificial*" bacteria and viruses, by starting with a synthesized "*minimal genome*" which was capable of self-replication (a kind of empty shell), to which "*desired*" traits could be added at will.
4. Creating hybrids of related viral strains.

These possibilities multiply as scientists begin publishing the complete DNA sequences of well-known pathogens: "... [G]enomics efforts in laboratories around the world will deliver the complete sequence of more than 70 major bacterial, fungal, and parasitic pathogens of humans, animals and plants in the next year or two...." Scientists working in these areas point out that actually getting engineered viruses and bacteria to survive in the environment, and to be maximally useful as weapons of war and terrorism, would not be easy to do; moreover, defenses against them can be constructed. What we are faced with the advances in molecular genetics, therefore, is an increase in the risks (possible harms) of novel agents being used in these ways for nefarious purposes.

There are *related risks* from accidental or unintended consequences of genomics research, especially from the genetic engineering of viruses and bacteria, which could result from the escape into the environment of virulent new organisms, irrespective of whether these organisms were intended originally for "*beneficent*" or "*malevolent*" purposes.

There was a brief flurry of publicity earlier this year when Australian researchers announced that, in engineering the relatively harmless mousepox virus with a gene for the chemical *interleukin 4*, in an attempt to create a contraceptive vaccine for mice, they had accidentally made the virus exceptionally toxic: "The virus does not directly threaten humans. But splice the IL-4 gene into a human virus and you could create a potent weapon. Add the gene to a pig virus, say, and you could wreck a nation's food supply".

There are risks to the "*nature*" of humans and other animals from intended or unintended consequences of genetic manipulations that either introduce reproducible changes into an existing genome (e.g., human or animal germ-line gene therapy), thus modifying existing species, or create entirely new variant species. For illustration, example of "*chimeras*", that is, combined entities made up of parts of the genome of two or more different species, including of course humans. Some molecular biologists apparently already have done casual experiments inserting human DNA into the eggs of other animals and growing the cell mass for a week or so; and there is much speculation as to what would happen if human and chimpanzee DNA were crossed, since chimps share over 98 % of human genes.

The DNA of all species now on earth is composed of the same four chemical bases, abbreviated A, T, C, G, arranged into two pairs (A/T, C/G), that make up the "ladders" on the double helix of DNA; different combinations of the base-pairs specify one of 20 amino acids, which combine to form various proteins. Some scientists are experimenting with adding more chemicals that would act as new bases, so that, for example, there would be six rather than four bases and

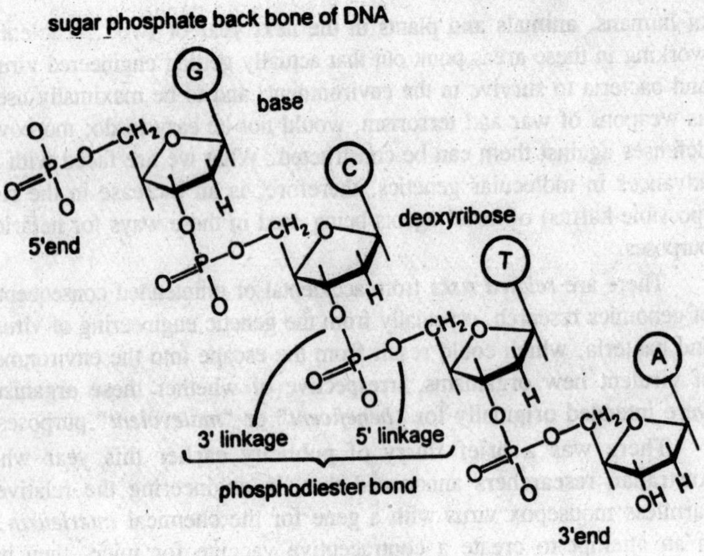

Fig. 6.3. Nitrogen base pairing in DNA.

perhaps three base-pairs. One of the scientists doing this work is Peter Schultz: "Schultz often says living things have only 20 amino acids because God rested on the seventh day. 'If He worked on Sunday,' he said, 'what would we look like?' The self-comparison between Dr. Schultz and God is interesting, to say the least.

There have been widely-publicized discussions of certain unique risks to organic life, stemming from possibilities allegedly inherent in the development of robotics and nanotechnology, especially in a now-infamous paper by Bill Joy, Chief Scientist at Sun Microsystems and creator of the "Java" script. Joy wrote:

The 21st-century technologies—genetics, nanotechnology, and robotics (GNR)—are so powerful that they can spawn whole new classes of accidents and abuses. Most dangerously, for the first time, these accidents and abuses are widely within the reach of individuals or small groups.... I think it is no exaggeration to say that we are on the cusp of the further perfection of extreme evil, an evil whose possibility spreads well beyond that which weapons of mass destruction bequeathed to the nation-states, on to a surprising and terrible empowerment of extreme individuals.

The link between *nanotechnology* and *biotechnology* is fascinating: Although the former works with intrinsically inert materials, it is seeking to turn them into a perfect analogue of a biological (self-

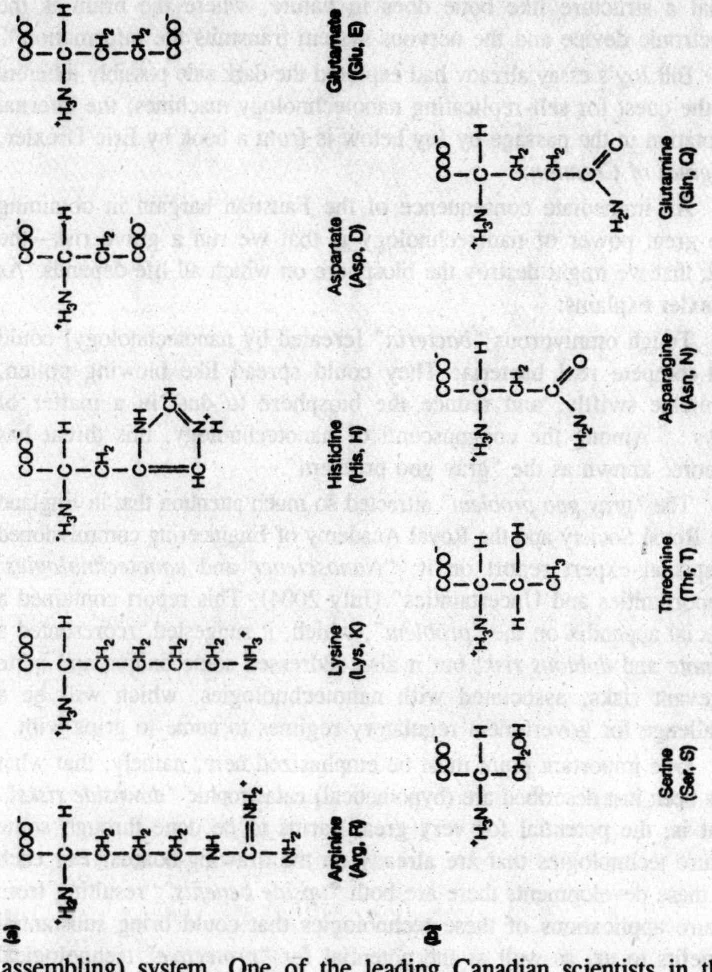

Fig. 6.4. The 20 standard amino acids.

assembling) system. One of the leading Canadian scientists in this field, Dragon Petrovic, has explained the quest as follows:

In the future, he predicts, technicians will teach individual molecules and atoms to assemble themselves into wires and sheets of impeccable purity and thinness.... instruments made of compounds that are self-assembled, atom by perfect atom—materials so pure that they could never snap apart or break under normal conditions.... "Imagine the linkage to telecom—can we get DNA molecules to self-assemble into perfect sheets and wires only an atom thick, and then send electrons and photons to stimulate the DNA to do things—start growing; stop growing; assemble into certain geometric shapes? It's analogous to what a structure like bone does in nature, where the brain is the electronic device and the nervous system transmits the information".

Bill Joy's essay already had explored the dark side possibly inherent in the quest for self-replicating nanotechnology machines; the internal quotation in the passage by Joy below is from a book by Eric Drexler, *Engines of Creation*:

An immediate consequence of the Faustian bargain in obtaining the great power of nanotechnology is that we run a grave risk—the risk that we might destroy the biosphere on which all life depends. As Drexler explains:

Tough omnivorous "*bacteria*" [created by nanotechnology] could out-compete real bacteria: They could spread like blowing pollen, replicate swiftly, and reduce the biosphere to dust in a matter of days.... Among the congnoscenti of nanotechnology, this threat has become known as the "gray goo problem".

The "*gray goo problem*" attracted so much attention that in England the Royal Society and the Royal Academy of Engineering commissioned a special expert report on it: "*Nanoscience* and *nanotechnologies*: Opportunities and Uncertainties" (July 2004). This report contained a special appendix on the "*problem*", which, it suggested, represented a *remote* and *dubious risk*; but it also addressed some unique and quite relevant risks, associated with nanotechnologies, which will be a challenge for government regulatory regimes to come to grips with.

One important point must be emphasized here, namely, that what has been just described are (hypothetical) catastrophic "*downside risks*", that is, the potential for very great harms to be done through some future technologies that are already on the drawing-boards. For each of these developments there are both "*upside benefits*", resulting from future applications of these technologies that could bring substantial benefits to us, as well as the potential for "*protective*" technological innovations that could mitigate, offset, reduce, or even eliminate at least some of the downside risks. To take the example of the engineering of viruses as bioweapons: As a counter to this threat (and also just to reduce the debilitating effects of viral infections on population health), research is under way in molecular genetics to develop new antiviral drugs that can block the infectious action of any viruses at the cellular level (preventing receptor binding, cell penetration, replication, production of viral proteins, and so on). Considered as a totality, however, what these conjoined prospects do is to continually "*raise the stakes*" in our technological game with nature, whereby the new sets of risks and benefits reflect both, and simultaneously, the potential for an upside of hitherto unattainable benefits and a downside of hitherto unimaginable horrors. As discussed in a later section, this entire prospect increases the challenge to our social institutions to manage our technological prowess so as to realize the benefits and avoid the harms, and likewise increases the risk that we will be unable to do so.

What is Different Today

There are undoubtedly other types of *catastrophic risks*, but those introduced above are sufficient for purposes of discussion! My main point is that these newer risks are fundamentally different in character from the case of nuclear winter, and the difference has to do with the distribution of knowledge and technological capacity relevant to them (thus requiring a very different institutional response). The technologies giving rise to the nuclear winter risk are controlled by just two nation-states and are maintained (for the most part, and until now) under a thick blanket of military security and secrecy, although the smuggling of nuclear materials out of the former Soviet Union is cause for worry. Both the essential theoretical knowledge, and the engineering capacity needed to turn that into weapons, is confined to a relatively small circle of experts and officials. Not so with the new technologies.

The catastrophic risk areas listed above stem from current research programs that are widely distributed around the world; moreover, the strongest drivers of them are private corporations, including the large pharmaceutical multinationals, acting with full encouragement, support, and incentives from national governments. Especially where the possible health benefits of genetic manipulations are concerned, the combined public-private interests are overwhelmingly supportive, driving the research ahead at an accelerating pace. Governments especially are enthralled with the economic significance of these new technologies, are competing with each other under innovation agendas to capture major shares of the corporate investments, and are loathe to stop and think about unintended consequences.

All of the characteristics of the knowledge and applications in these areas mean that it is extremely difficult even to think about controlling either the process or the results. For one thing, the knowledge is widely distributed among individual scientists; for another, it is widely distributed among private actors (corporations) which have the option of moving their operations on a regular basis, seeking perhaps the least-regulatory-intensive national base on the globe. Third, the technologies themselves become increasingly "*simplified*" and thus easier to hide, if necessary; the genetics technologies, for example, can be carried out in small laboratories almost anywhere. Sergei Popov, the Russian scientist who pioneered germ warfare research using recombinant DNA techniques, observed recently: "The whole technology becomes more and more available. It becomes easier and easier to create new biological entities, and they could be quite dangerous".

Fourth, oversight is inhibited by the lure of truly extraordinary economic and health benefits promised by the new knowledge and technologies. And fifth, just the astonishing pace of innovation itself today makes the prospect of control and regulation a challenge.

During the past year national governments have been scrambling to respond to just a few of the dimensions of these new risks. Most attention has been focused on human cloning, where a few rogue scientists have challenged authorities in various jurisdictions to "*try to stop us*", and laws prohibiting this technology are being passed rapidly. But this is a relatively crude technology, albeit one which excites public attention, and one wonders whether authorities will become complacent about their ability to control unacceptable technologies due to their experience with this case. (Meanwhile, there are increasing reports that many genetics scientists are "going underground", in the sense that they have stopped talking publicly about their research in progress for fear that public reactions will be hostile and will result in official steps to halt it.)

Bill Joy and Ian Ramshaw have called for urgent action under the Biological and Toxic Weapons Convention (1975, hereafter BTWC), to provide explicitly for a global oversight effort over some of the new technologies and their applications described earlier. Unfortunately, and ironically in view of what was to happen only two months later, at a meeting of the parties in Australia in July 2001 the United States unexpectedly blocked the process of completing a protocol under the BTWC that would have made the Convention something other than a statement of good intentions, for in its present form it has no provisions for verification or compliance monitoring. The US government has been pressured by its biotechnology industry sector not to agree to a verification protocol, under which inspections of laboratories and other facilities by international teams of experts would be carried out in all the signatory countries, because industry fears that its intellectual property and commercial secrets could be compromised. At the time of writing other signatories were considering whether they should proceed to complete the adoption of the verification protocol without US support.

Unfortunately, we know international negotiation to be at the best of times a tedious and protracted process, and there is reason to believe that in this domain it could be fractious and unsuccessful. This is because all of the technologies described represent frontiers of industrial innovation in which great multinational corporations and the national governments which protect their interests have significant investments; both corporations and governments would be loathe to see those investments and the immense payoffs expected from them jeopardized by an international control regime. A recent article co-authored by a molecular geneticist and a specialist in the international convention on biological weapons has called for an urgent new effort to strengthen verification under the 1975 Convention and to enlist the biomedical research community in an effort to strengthen deterrence against the uses of bioengineered organisms for war and terrorism.

CONCLUSION

Now is the time for intensive exploration of the theme of policing science and to ask the following types of questions:

1. Can we characterize a set of new *catastrophic risks*, as defined here, related to the leading-edge technologies that are being developed?
2. Do these new risks have an essential character that will make them difficult to control, because the knowledge and the technologies will be so widely diffused?
3. Can these risks be confined to acceptable dimensions by the institutional means now at our disposal, including international conventions on prohibitions? If not, what new tools do we need, and how can we get them?
4. Do professional associations of scientists working in these fields have special responsibilities to assist societies in controlling these risks, and if so, are those responsibilities now being discharged adequately?

What is at risk in this game, now, is the possibility that the tension between science and society will become both unmanageable for institutions and unbearable for individuals, in other words, that the destructive applications of our operational power finally will overwhelm the rest. This possibility arises out of the striking contrast between the pace of change in social and legal institutions (especially international agreements), on the one hand, and in new scientific and technological breakthroughs in the sciences, especially in genomics, including applications relevant to biowarfare and bioterrorism on the other. In the first-mentioned the pace is painfully slow and progress often remains ineffective even after decades of negotiation, as in the case of the Convention on Biological and Toxic Weapons. The second proceeds at a frenetic and steadily-accelerating pace.

To reduce the probability that change in the second will overwhelm our social and legal capacity to steer technological development away from the zone of catastrophic risks, it is necessary first to get agreement among influential social actors that this is, as described here, a momentous challenge which contemporary society cannot avoid. The first practical test of our resolve in this regard, is whether influential scientists can be mobilized in the cause, scientists who will reaffirm the need for new oversight structures, to be erected both within the practice of science itself and also in the relation between science and society. Hegel made a remark, somewhere in his writings, to the effect that only the hand which inflicts a wound can heal it. The wound here is the rupture with the dominant pre-modern relation of humanity and nature, governed by value-laden categories of being, and its replacement by modern science's purely operational orientation to the totality of the natural world.

Not in any "*ontological*" sense, but in a practical sense, as a matter of public policy, it is clear what is required —namely, that the practitioners of science join others in a program to try to bring our operational powers under the control and direction of social institutions that have universal validity, ones that correspond in sufficient measure with the common aspirations of humanity. It is my contention that today's dominant institutions do not have such validity and that, as a result, everyone on earth is at risk of having these powers become instruments in an Armageddon waged to the bitter end by contending social, ethnic, national, and religious interests.

What remains to be seen is whether the task as defined here can be widely recognized and grasped as such, while there is still time, and whether our scientific enterprise can be steered towards the shelter of a social compact having universal validity. If it turns out that despite our best efforts this cannot be done, there will arise a set of other questions that, for now at least, are too abhorrent for many even to consider. These questions have to do with the possibility that, taking both "normal" human passions and human institutional failings into consideration, there may be forms of knowledge that, as a practical matter, are too dangerous for us to possess, and that our only choice is to renounce and suppress such knowledge or suffer the consequences. In mentioning them we go to the heart of the fateful compact between science and society that has set the course for the development of modern society from the seventeenth century onwards, under the program known as the domination of nature. It is likely that contemporary society is not ready to deal with them, at least, not yet.

7

ART OF BUILDING SMALL

Make it small!" is a technological edict that has changed the world. The development of *microelectronics*—first the *transistor* and then the aggregation of transistors into *microprocessors*, *memory chips* and controllers—has brought forth a cornucopia of machines that manipulate information by streaming electrons through silicon. The idea of making "*nanostructures*" that comprise just one or a few atoms has great appeal, both as a scientific challenge and for practical reasons. A structure the size of an atom represents a fundamental limit: to make anything smaller would require manipulating atomic nuclei—essentially, transmuting one chemical element into another. In recent years, scientists have learned various techniques for building nanostructures, but they have only just begun to investigate their properties and potential applications. The age of *nanofabrication* is here, and the age of nanoscience has dawned, but the age of nanotechnology—finding practical uses for nanostructures—has not really started yet.

Researchers may well develop nanostructures as *electronic components*, but the most important applications could be quite different: for example, biologists might use nanometer-scale particles as minuscule sensors to investigate cells. Because scientists do not know what kinds of nanostructures they will ultimately want to build, they have not yet determined the best ways to construct them. Photolithography, the technology used to manufacture computer chips and virtually all other microelectronic systems, can be refined to make structures smaller than 100 nanometers, but doing so is very difficult, expensive and inconvenient. In a search to find better alternatives, nanofabrication researchers have adopted the philosophy "*Let a thousand flowers bloom.*"

First, consider the advantages and disadvantages of *photolithography*. Manufacturers use this phenomenally productive technology to churn out three billion transistors *per second* in the U.S. alone. *Photolithography* is basically an extension of photography. One first makes the equivalent of a photographic negative containing the pattern required for Some part of a microchip's circuitry This negative, which is called the *mask* or *master*, is then used to copy the pattern into the metals and semiconductors of a *microchip*. As is the case with photography, the negative may be hard to make, hut creating multiple copies is easy, because the mask can be used many times. The process thus separates into two stages: the preparation of the mask (a one-time event, which can he slow and expensive) and the use of the mask to manufacture replicas (which must be rapid and inexpensive).

To make a mask for a part of a *computer chip*, a manufacturer first designs the circuitry pattern on a conveniently large scale and converts it into a pattern of opaque metallic film (usually chromium) on a transparent plate (usually glass or silica). Photolithography then reduces the sire of the pattern in a process analogous to that used in a photographic darkroom. A beam of light (typically ultraviolet light from a mercury-arc lamp) shines through the chromium mask, then passes through a lens that focuses the image onto a photosensitive coating of organic polymer (called the *photoresist*) on the surface of a silicon wafer. The parts of the photoresist struck by the light can be selectively removed, exposing parts of the silicon wafer in a way that replicates the original pattern.

Why not use photolithography to make *nanostructures*? The technology faces two limitations. The first is that the shortest wavelength of ultraviolet light currently used in production processes is about 250 nanometers. Trying to make structures much smaller than half of that spacing is like trying to read print that is too tiny: diffraction causes the features to blur and melt together. Various technical improvements have made it possible to push the limits of photolithography. The smallest structures created in mass production are somewhat larger than too nanometers, and complex microelectronic structures have been made with features that are only 70 nanometers across. But these structures are still not small enough to explore some of the most interesting aspects of nanoscience.

The second limitation follows from the first: because it is technically difficult to make such small structures using light, it is also very expensive to do so. The *photolithographic tools* that will be used to make chips with features well below 100 nanometers will each cost tens to hundreds of millions of dollars. This expense may or may not be acceptable to manufacturers, but it is prohibitive for the biologists, materials scientists, chemists and physicists who wish to explore nanoscience using structures of their own design.

Electronic Industry

The *electronics industry* is deeply interested in developing new methods for *nanofabrication* so that it can continue its long-term trend of building ever smaller, faster and less expensive devices, It would he a natural evolution of microelectronics to become nanoelectronics. But because conventional photolithography becomes more difficult as the dimensions of the structures become smaller, manufacturers are exploring alter native technologies for making future nanochips.

One leading contender is electron-beam lithography. In this method, the circuitry pattern is written on a thin polymer film with a beam of electrons. An electron beam does not diffract at atomic scales, so it does not cause blurring of the edges of features. Researchers have used the technique to write lines with widths of only a few nanometers in a layer of photoresist on a silicon substrate.

The *electron-beam instruments* currently available, however, are very expensive and impractical for large-scale manufacturing. Because the beam of electrons is needed to fabricate each structure, the process is similar to the copying of a manuscript by hand, one line at a time.

If electrons are not the answer, what is? Another contender is lithography using x-rays with wavelengths between 0.1 and 10 nanometers or extreme ultraviolet light with wavelengths between 10 and 70 nanometers. Because these forms of radiation have much shorter wavelengths than the ultraviolet light currently used in photolithography, they minimize the blurring caused by diffraction. These technologies face their own set of problems however: conventional lenses are not transparent extreme ultraviolet light and do not focus x-rays. In the energetic radiation rapidly damages many of the materials used in masks and lenses. But the microelectronics industry clearly would prefer to make advanced chips using extensions of familiar technology, so these methods are being actively developed. Some of the techniques (for example, advanced ultraviolet lithography for chip production) will probably become commercial realities. They will not, though, make inexpensive nanostructures and thus will do nothing to open nanotechnology to a broader group of scientists and engineers.

The need for simpler and less expensive methods of *fabricating nanostructures* has stimulated the search for unconventional approaches that have not been explored by the electronics industry. We first became interested in the topic in the 1990s when we were engaged in making the simple structures required in microfluidic systems—chips with channels and chambers for holding liquids. This lab-on-a-chip has myriad potential uses in biochemistry, ranging from drug screening to genetic analysis. The channels in microfluidic chips are enormous by the standards of microelectronics: 50 microns (or 50,000 nanometers) wide, rather than 100 nanometers. But the techniques for producing those channels are quite versa tile. *Microfluidic chips* can be made quickly and inexpensively, and many are composed of organic polymers and gels—materials not bond in the world of electronics. We discovered that we could use similar techniques to create nanostructures.

The methods represented, in a sense, a step backward in technology. Instead of using the tools of physics—light and electrons—we employed mechanical processes that are familiar in everyday life: printing, stamping, molding and embossing. The techniques are called soft lithography because the tool they have in common is a block of *polydimethylsiboxane* (PDMS)—the rubbery polymer used to caulk the leaks around bathtubs. (Physicists often refer to such organic chemicals as "soft matter.")

Lithography

To carry out reproduction using *soft lithography*, one first makes a mold or a stamp. The most prevalent procedure is to use photolithography or electron-beam lithography to produce a pattern in a layer of photoresist on the surface of a silicon wafer, This process generates a bas-relief master in which islands of photoresist stand out from the silicon. Then a chemical precursor to PDMS—a free-flowing liquid—is poured over the bas-relief master and cured into the rubbery solid. The result is a PDMS stamp that matches the original pattern with astonishing fidelity: the stamp reproduces features from the master as small as a few nanometers. Although the creation of a finely detailed has-relief master is expensive because it requires electron-beam lithography or other advanced techniques, copying the pattern on PDMS stamps is cheap and easy And once a stamp is in hand, it can be used in various inexpensive ways to make nanostructures.

The first method—original developed by Amit Kumar, a postdoctoral student in our group at Harvard University—is called *microcontact printing*. The PDMS stamp is "*inked*" with a reagent solution consisting of organic molecules called *thiols*. The stamp is then brought into contact with an appropriate sheet of "*paper*"—a thin film of gold on a glass, silicon or polymer plate. The thiols react with the gold surface, forming a highly ordered film (called a *self monolayer*, or *SAM*) that replicates the stamp's pattern. Because the thiol ink spreads a bit after it contacts the surface, the resolution of the monolayer cannot be quite as high as that of the PDMS stamp. But when used correctly, micro contact printing can produce patterns with features as small as 50 nanometers.

Another method of soft lithography, called *micromolding* in capillaries involves using the PDMS stamp to mold patterns. The stamp is placed on a hard surface, and a liquid polymer flows by capillary action into the recesses between the surface and the stamp. The polymer then solidifies into the desired pattern. This technique can replicate structures smaller than 10 nanometers. It is particularly well suited for producing sub-wavelength optical devices, waveguides and optical polarizer all of which could be used in optical fiber networks and eventually perhaps in optical computers. Other possible applications are in the field of nanofluidics, an extension of microfluidics that would involve producing chips for biochemical research with channels only a few nanometers wide. At that scale, fluid dynamics may allow new ways to separate materials such as fragments of DNA.

These methods require no special equipment and in fact can he carried out by hand in an ordinary laboratory. Conventional photolithography must take place in a clean-room facility devoid of dust and dirt; if a piece of dust lands on the mask, it will create an unwanted spot on the pattern. As a result, the device being fabricated (and sometimes neighbouring devices) may fail. Soft lithography is generally more forgiving because the PDMS stamp is elastic. If a piece of dust gets trapped between the stamp and the surface, the stamp will compress over the top of the particle but maintain contact with the rest of the surface. Thus, the pattern will he reproduced correctly except for where the contaminant is trapped.

Moreover, soft lithography can produce nanostructures in a wide range of materials, including the complex organic molecules needed for biological studies. And the technique can print or mold patterns on curved as well as planar surfaces. But the technology is not ideal for making the structures required for complex nanoelectronics. Currently all integrated circuits consist of stacked layers of different materials. Deformations and distortions of the soft PDMS stamp can produce small errors in the replicated pattern and a misalignment of the pat tern with any underlying patterns previously fabricated. Even the tiniest distortions or misalignments can destroy a multilayered nano electronic device. Therefore, soft lithography is not well suited for fabricating structures with multiple layers that must stack precisely on top of one another.

Researchers have found ways, however, to correct this shortcoming—at least in part—by employing a rigid stamp instead of an elastic one. In a technique called *step-and-flash* imprint lithography, developed by C. Grant Willson of the University of Texas, photolithography is used to etch a pattern into a quartz plate, yielding a rigid bas-relief master. Willson eliminated the step of making a PDMS stamp from the master; instead the master itself is pressed against a thin film of liquid polymer, which fills the master's recesses. Then the master is exposed to ultraviolet light, which solidifies the polymer to create the desired replica. A related technique called *nanoimprint lithography*, developed by Stephen Y. Chou of Princeton University, also employ a rigid master but uses a film of polymer that has been heated to a temperature near its melting point to facilitate the embossing process. Both methods can produce two-dimensional structures with good fidelity, but it remains to be seen whether the techniques are suitable for manufacturing electronic devices.

The current revolution in nano science started in 1981 with the invention of the *scanning tunneling microscope* (STM), for which Heinrich Rohrer and Gerd K. Binnig of the IBM Zurich Research Laboratory received the Nobel Prize in Physics in 1986. This remarkable device detects small currents that pass between the microscope's tip

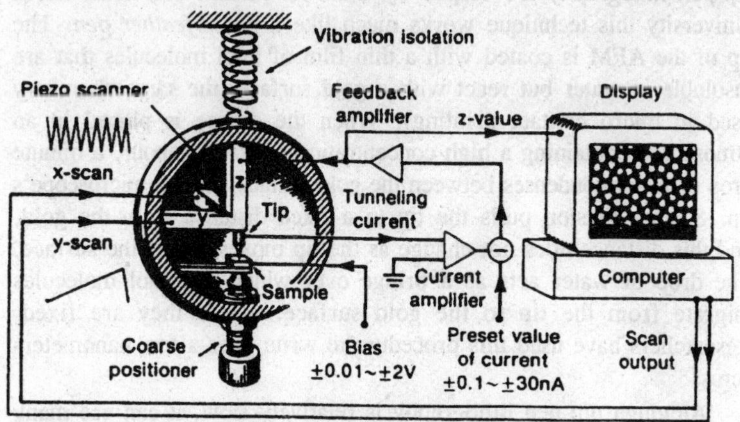

Fig. 7.1. Scanning tunneling microscope (STM).

and the sample being observed, allowing researchers to "see" substances at the scale of individual atoms, The success of the STM led to the development of other scanning probe devices, including the atomic force microscope (AFM). The operating principle of the AFM is similar to that of an old-fashioned phonograph. A tiny probe—a fiber or a pyramid-shaped tip that is typically between two and 30 nanometers wide—is brought into direct contact with the sample. The probe is attached to the end of a cantilever, which bends as the tip moves across the sample's surface. The deflection is measured by reflecting a beam of laser light off the top of the cantilever. The AFM can detect variations in vertical surface topography that are smaller than the dimensions of the probe.

But scanning probe devices can do more than simply allow scientists to observe the atomic world—they can also be used to create nanostructures. The tip on the AFM can be used to physically move nano particles around on surfaces and to arrange them in patterns. It can also he used to make scratches in a surface (or more commonly, in monolayer films of atoms or molecules that coat the surface). Similarly, if researchers increase the currents flowing from the tip of the STM, the microscope becomes a very small source for an electron beam, which can be used to write nanometer-scale patterns. The STM tip can also push individual atoms around on a surface to build rings and wires that are only one atom wide.

Dip-pen Lithography

An intriguing new scanning probe fabrication method is called *dip-pen lithography* Developed by Chad A. Mirkin of Northwestern University this technique works much like a *goose-feather pen*. The tip of the AFM is coated with a thin film of thiol molecules that are insoluble in water but react with a gold surface (the same chemistry used in micro contact printing). When the device is placed in an atmosphere containing a high concentration of water vapour, a minute drop of water condenses between the gold surface and the microscope's tip. Surface tension pulls the tip to a fixed distance from the gold, and this distance does not change as the tip moves across the surface. The drop of water acts as a bridge over which the thiol molecules migrate from the tip to the gold surface, where they are fixed. Researchers have used this procedure to write lines a few nanometers across.

Although dip-pen lithography is relatively slow, it can use many different types of molecules as "*inks*" and thus brings great chemical flexibility to nanometer-scale writing. Researchers have not yet determined the best applications for the technique, but one idea is to use the dip-pen method for precise modifications of circuit designs. Mirkin has recently demonstrated that a variant of the ink used in dip-pen lithography can write directly on silicon.

An interesting cousin to these techniques involves another kind of nanostructure, called a *break junction*. If you break a thin, ductile metal wire into two parts by pulling sharply, the process seems abrupt to a human observer, but it actually follows a complex sequence. When the force used in breaking the wire is first applied, the metal begins to yield and flow, and the diameter of the wire decreases As the two ends move apart, the wire gets thinner and thinner until, in the instant just before breaking, it is a single atom in diameter at its narrowest point. This process of thinning a wire to a break junction can he detected easily by measuring the current that flows through the wire. When the wire is slender enough, current can flow only in discrete quantities (that is, current flow is quantized).

The *break junction* is analogous to two STM tips facing each other, and similar physical rules govern the current that flows through it. Mark A. Reed of Yale University has pioneered a particularly inventive use of the break junction. He built a device that enabled a thin junction to be broken under carefully controlled conditions and then allowed the broken tips to be brought back together or to be held apart at any distance with an accuracy of a few thousandths of a nanometer. By adjusting the distance between the tips in the presence of an organic molecule that bridged them, Reed was able to measure a cur rent flowing across the organic bridge. This experiment was an important step in the development of technologies for using single organic molecules as electronic devices such as diodes and transistors.

All the forms of lithography we have discussed so far are called *top-down methods*—that is, they begin with a pattern generated on a larger scale and reduce its lateral dimensions (often by a factor of 10) before carving out nanostructures This strategy is required in fabricating electronic devices such as microchips, whose functions depend more on their patterns than on their dimensions. But no top-down method is ideal; none can conveniently, cheaply and quickly make nanostructures of any material. So researchers have shown growing interest in bottom-up methods, which start with atoms or molecules and build up to nanostructure These methods can easily make the smallest nano-structures—with dimensions between two and 10 nanometers—and do so inexpensively. But these structures are usually generated as simple particles in suspension or on surfaces, rather than as designed, interconnected patterns.

Two of the most prominent *bottom-up methods* are, those used to make nanotubes and quantum dots. Scientists have made long, cylindrical tubes of carbon by a catalytic growth process that employs a nanometer-scale drop of molten metal (usually iron) as a catalyst. The most active area of research in quantum dots originated in the laboratory of Louis E. Brus and has been developed by A. Paul Alivisatos of the University of California at Berkeley, Moungi G. Bawendi of the Massachusetts institute of Technology, and others. Quantum dots are crystals containing only a few hundred atoms. Because the electrons in a quantum dot are con fined to widely separated energy levels, the dot emits only one wavelength of light when it is excited. This property makes the quantum dot useful as a biological marker.

One procedure used to make *quantum dots* involves a chemical reaction between a metal ion (for example, cadmium) and a molecule that is able to donate a selenium ion. This reaction generates crystals of cadmium selenide. The trick is to prevent the small crystals from sticking together as they grow to the desired size. To insulate the growing particles from one another, researchers carry out the reaction in the presence of organic molecules that act as surfactants, coating the surface of each cadmium selenide particle as it grows. The organic molecules stop the crystals from clumping together and regulate their rate of growth. The geometry of the particles can be controlled to some extent by mixing different ratios of the organic molecules. The reaction can generate particles with a variety of shapes, including spheres, rods and tetrapods (four-armed particles similar to toy jacks).

It is important to synthesize the quantum dots with uniform size and composition, because the size of the dot determines its electronic, magnetic and optical properties. Researchers can select the size of the particles by varying the length of time for the reaction. The organic coating also helps to set the size of the particles. When the nanoparticle is small (on the scale of molecules), the organic coating is loose and allows further growth; as the particle enlarges, the organic molecules become crowded. There is an optimum size for the particles that allows the most stable packing of the organic molecules and thus provides the greatest stabilization for the surfaces of the crystals.

These cadmium selenide nanoparticles promise some of the first commercial products of nanoscience: *Quantum Dot Corporation* has been developing the crystals for use as biological labels. Researchers can tag proteins and nucleic acids with quantum dots; when the sample is illuminated with ultraviolet light, the crystals will fluoresce at a specific wavelength and thus show the locations of the attached proteins. Many organic molecules also fluoresce, but quantum dots have several advantages that make them better markers. First, the colour of a quantum dot's fluorescence can be tailored by changing the dot's size: the larger the particle the more the emitted light is shifted toward the red end of the spectrum. Second, if all the dots are the same size, their fluorescence spectrum is narrow—that is, they emit a very pure colour. This property is important because it allows particles of different sizes to be used as distinguishable labels. Third, the fluorescence of quantum dots does not fade on exposure to ultraviolet light, as does that of organic molecules. When used as dyes in biological research, the dots can be observed for conveniently long periods.

Scientists are also investigating the possibility of making structures from colloids—nanoparticles in suspension. Christopher B. Murray and a team at the IBM Thomas J. Watson Research Center are exploring the use of such colloids create a medium for ultrahigh-density data storage. The IBM team's colloids contain magnetic nanoparticles as small as three nanometers across, each composed of about 1,000 iron and platinum atoms. When the colloid is spread on a surface and the solvent allowed to evaporate, the nanoparticles crystallize in two- or three-dimensional arrays. Initial studies indicate that these arrays can potentially store trillions of bits of data per square inch, giving them a capacity 10 to 100 times greater than that of present memory devices.

The interest in *nanostructures* is so great that every plausible fabrication technique is being examined. Although physicists and chemists are now doing most of the work, biologists may also make important contributions. The cell (whether mammalian or bacterial) is relatively large on the scale of nanostructures: the typical bacterium is approximately 1,000 nanometers long, and mammalian cells are larger. Cells are, however, filled with much smaller structures, many of which are astonishingly sophisticated. The ribosome, for example, carries out one of the most important cellular functions: the synthesis of proteins from amino acids, using messenger RNA as the template. The complexity of this molecular construction project far surpasses that of man-made techniques. Or consider the rotary motors of the bacterial flagella, which efficiently propel the one-celled organisms.

It is unclear if "*nanomachines*" taken from cells will be useful. They will probably have very limited application in electronics, but they may provide valuable tools for chemical synthesis and sensing devices. Recent work by Carlo D. Montemagno of Cornell University has shown that it is possible to engineer a primitive nano machine with a biological engine. Montemagno extracted a rotary motor protein from a bacterial cell and connected it to a metallic nanorod—a cylinder 750 nanometers long and 150 nanometers wide that had been fabricated by lithography. The rotary motor, which was only 11 nanometers tall, was powered by *adenosine triphosphate* (ATP), the source of chemical energy in cells. Montemagno showed that the motor could rotate the nano rod at eight revolutions per minute. At the very least, such research stimulates efforts to fabricate functional nanostructure by demonstrating that such structures can exist.

The development of *nanotechnology* depend on the avail ability of nanostructures. The invention of the STM and AFM has provided new tools for viewing, characterizing and manipulating these structures; the issue now is how to build them to order and how to design them to have new and useful functions. The importance of electronics applications has tended to focus attention on nanodevices that might he incorporated into future integrated circuits. And for good technological reasons, the electronics industry has emphasized fabrication methods that are extensions of those currently used to make microchips. But the explosion of interest in nanoscience has created a demand for a broad range of fabrication methods, with an emphasis on low-cost, convenient techniques

The new approaches to nanofabrication are unconventional only because they are not derived from the microtechnology developed for electronic devices. Chemists, physicists and biologists are rapidly accepting these techniques as the most appropriate ways to build various kinds of nanostructures for research. And the methods may even supplement the conventional approaches—photolithography, electron-beam lithography and related techniques—for applications in electronics as well. The microelectronics mold is now broken. Ideas for nano-fabrication are coming from many directions in a wonderful free-for-all of discovery.

8

LONG LIFE IN AN OPEN WORLD

Cell repair machines raise questions involving the value of extending human life. These are not the questions of today's medical ethics, which commonly involve dilemmas posed by scarce, costly, and half-effective treatments. They are instead questions involving the value of long, healthy lives achieved by inexpensive means. For people who value human life and enjoy living, such question may need no answer. But after a decade marked by concern about population growth, pollution, and resource depletion, many people may question the desirability of extending life; such concerns have fostered the spread of pro-death memes. These memes must be examined afresh, because many have roots in an obsolete worldview. Nanotechnology will change far more than just human life-span.

We will gain the means not only to heal ourselves, but to heal Earth of the wounds we have inflicted. Since saving lives will increase the number of the living, life extension raises questions about the effect of more people. Our ability to heal the Earth will lessen one cause for controversy.

Still, *cell repair machines* themselves will surely stir controversy. They disturb traditional assumptions about our bodies and our futures: this makes doubt soothing. They will require several major break-throughs: this makes doubt easy. Since the possibility or impossibility of cell repair machines raises important issues, it makes sense to consider what objections might be raised.

CELL REPAIR MACHINES

What sort of argument could suggest that cell repair machines are impossible? A successful argument must manage some strange contortions. It must somehow hold that molecular machines cannot build and repair cells, while granting that the molecular machines in our bodies actually do build and repair cells every day. A cruel problem for the committed skeptic! True, artificial machines must do what natural machines fail to do, but they need not do anything *qualitatively* novel. Both natural and artificial repair devices must reach, identify, and rebuild molecular structures. We will be able to improve on existing DNA repair enzymes simply by comparing several DNA strands at once, so nature obviously hasn't found all the tricks. Since this example explodes any general argument that repair machines cannot improve on nature, a good case against cell repair machines seems difficult to make.

Still, two general questions deserve direct answers. First, why should we expect to achieve long life in the coming decades, when people have tried and failed for millennia? Second, if we can indeed use cell repair machines to extend lives, then why hasn't nature (which has been repairing cells for billions of years) already perfected them?

Hit and Trial

For centuries, people have longed to escape their short life spans. Every so often, a Ponce de Leon or a quack doctor has promised a potion, but it has never worked. These statistics of failure have persuaded some people that, since all attempts have failed, all always will fail. They say "*Aging is natural*," and to them that seems reason enough. Medical advances may have shaken their views, but advances have chiefly reduced early death, not extended maximum life-span.

But now biochemists have gone to work examining the machines that build, repair, and control cells. They have learned to assemble viruses and reprogram bacteria. For the first time in history, people are examining their molecules and unraveling the molecular secrets of life. It seems that *molecular engineers* will eventually combine improved biochemical knowledge with improved molecular machines, learning to repair damaged tissue structures and so rejuvenate them. This is nothing strange—it would be strange, rather, if such powerful knowledge and abilities did not bring dramatic results. The massive statistics of past failure are simply irrelevant, because we have never before tried to build cell repair machines.

Natural Efforts

Nature has been building *cell repair machines*. Evolution has tinkered with multi celled animals for hundreds of millions of years, yet advanced animals all age and die, because *nature's nanomachines* repair cells imperfectly. Why should improvements be possible?

Rats mature in months, and then age and die in about two years—yet human beings have evolved to live over thirty times longer. If longer lives were the chief goal of evolution, then rats would live longer too. But durability has costs: to repair cells requires an investment in energy, materials, and repair machines. Rat genes direct rat bodies to invest in swift growth and reproduction, not in meticulous self-repair. A rat that dallied in reaching breeding size would run a greater risk of becoming a cat snack first. Rat genes have prospered by treating rat bodies as cheap throwaways. Human genes likewise discard human beings, though after a life a few dozen times longer than a rat's.

But shoddy repairs are not the only cause of aging. Genes turn egg cells into adults through a pattern of development which rolls for ward at fairly steady speed. This pattern is fairly consistent because evolution seldom changes a basic design. Just as the basic pattern of the DNA-RNA-protein system froze several billions of years ago, so the basic pattern of chemical signals and tissue responses that guides

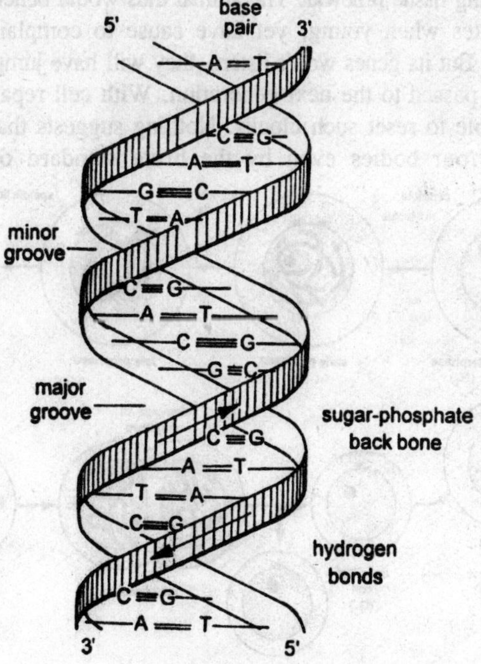

Fig. 8.1. Structure of DNA.

mammalian development jelled many millions of years ago. That process apparently has a clock, set to run at different speeds in different species, and a program that runs out.

Whatever the causes of aging, evolution has had little reason to eliminate them. If genes built individuals able to stay healthy for millennia, they would gain little advantage in their "*effort*" to replicate. Most individuals would still die young from starvation, predation, accident, or disease. As Sir Peter Medawar points out, a gene that helps the young (who are many) but harms the old (who are few) will replicate well and so spread through the population. If enough such genes accumulate, animals become programmed to die.

Experiments by Dr. Leonard Hay flick suggest that cells contain "*clocks*" that count cell divisions and stop the division process when the count gets too high. A mechanism of this sort can help young animals: if cancer-like changes make a cell divide too rapidly, but fail to destroy its clock, then it will grow to a tumour of limited size. The clock would thus prevent the unlimited growth of a true cancer. Such clocks could harm older animals by stopping the division of normal cells, ending tissue renewal. The animal thus would benefit from reduced cancer rates when young, yet have cause to complain if it lives to grow old. But its genes won't listen—they will have jumped ship earlier, as copies passed to the next generation. With cell repair machines we will be able to reset such clocks. Nothing suggests that evolution has perfected our bodies even by the brute standard of survival and

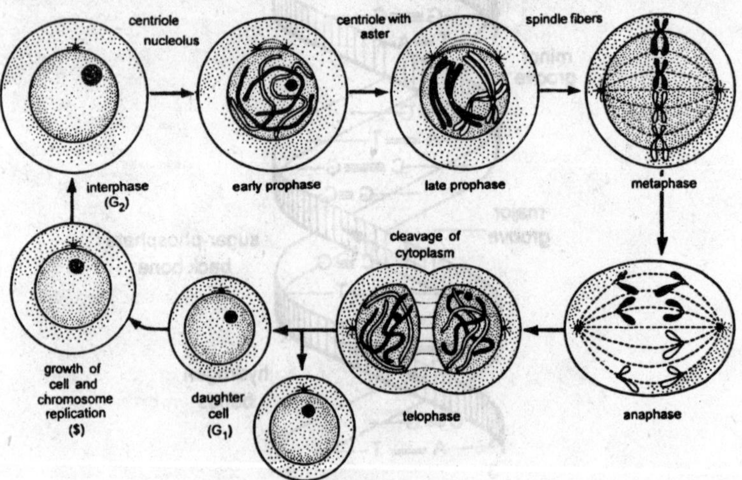

Fig. 8.2. Cell division.

reproduction. Engineers don't wire computers with slow, nerve like fibers or build machines out of soft protein, and for good reason. Genetic evolution (unlike memetic evolution) has been unable to leap to new materials or new systems, but has instead refined and extended the old ones.

The cell's repair machines fall far short of the limits of the possible—they don't even have computers to direct them. The lack of nanocomputers in cells, of course, shows only that computers couldn't (or simply didn't) evolve gradually from other molecular machines. Nature has failed to build the best possible cell repair machines, but there have been ample reasons.

Healing and Protecting the Earth

The failure of Earth's biological systems to adapt to the industrial revolution is also easy to understand. From deforestation to dioxin, we have caused damage faster than evolution can respond. As we have sought more food, goods, and services, our use of bulk technology has forced us to continue such damage. With future technology, though, we will be able to do more good for ourselves, yet do less harm to the Earth. In addition, we will be able to build planet-mending machines to correct damage already done. Cells are not all we will want to repair.

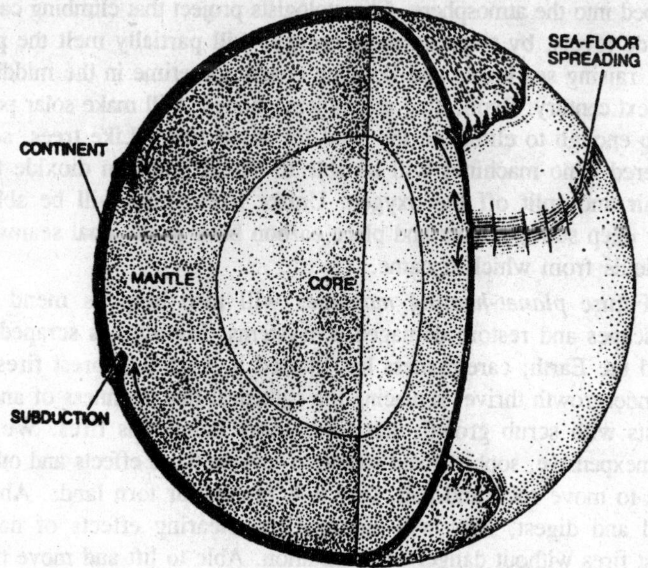

Fig. 8.3. The earth.

Consider the toxic waste problem. Whether in our air, soil, or water, wastes concern us because they can harm living systems. But any materials that come in contact with the molecular machinery of life can themselves be reached by other forms of molecular machinery. This means that we will be able to design cleaning machines to remove these poisons wherever they could harm life.

Some wastes, such as dioxin, consist of dangerous molecules made of innocuous atoms. Cleaning machines will render them harmless by rearranging their atoms. Other wastes, such as lead and radioactive isotopes, contain dangerous atoms. Cleaning machines will collect these for disposal in any one of several ways. Lead comes from Earth's rocks; assemblers could build it into rocks in the mines from which it came. Radioactive isotopes could also be isolated from living things, either by building them into stable rock or by more drastic means. Using cheap, reliable space transportation systems, we could bury them in the dead, dry rock of the Moon. Using nano machines, we could seal them in self-repairing, self-sealing containers the size of hills and powered by desert sunlight. These would be more secure than any passive rock or cask.

With replicating assemblers, we will even be able to remove the billions of tons of carbon dioxide that our fuel-burning civilization has dumped into the atmosphere. Climatologists project that climbing carbon dioxide levels, by trapping solar energy, will partially melt the polar caps, raising sea levels and flooding coasts sometime in the middle of the next century. Replicating assemblers, though, will make solar power cheap enough to eliminate the need for fossil fuels. Like trees, solar-powered nano machines will be able to extract car bon dioxide from the air and split off the oxygen. Unlike trees, they will be able to grow deep storage roots and place carbon back in the coal seams and oil fields from which it came.

Future *planet-healing machines* will also help us mend torn landscapes and restore damaged ecosystems. Mining has scraped and pitted the Earth; carelessness has littered it. Fighting forest fires has let undergrowth thrive, replacing the cathedral-like openness of ancient forests with scrub growth that feeds more dangerous fires. We will use inexpensive, sophisticated robots to reverse these effects and others. Able to move rock and soil, they will re-contour torn lands. Able to weed and digest, they will simulate the clearing effects of natural forest fires without danger or devastation. Able to lift and move trees, they will thin thick stands and reforest bare hills. We will make squirrel-sized devices with a taste for old trash. We will make treelike devices with roots that spread deep and cleanse the soil of pesticides and excess acid. We will make insect-sized lichen cleaners and spray-paint nibblers. We will make whatever devices we need to clean up the mess left by twentieth-century civilization.

After the cleanup, we will recycle most of these machines, keeping only those we still need to protect the environment from a cleaner civilization based on molecular technology. These more lasting devices will supplement natural ecosystems wherever needed, to balance and heal the effects of humanity. To make them effective, harmless, and hidden will be a craft requiring not just automated engineering, but knowledge of nature and a sense of art.

With *cell repair technology*, we will even be able to return some species from apparent *extinction*. The *African quagga*—a zebralike animal—became extinct over a century ago, but a salt-preserved quagga pelt survived in a German museum. Alan Wilson of the University of California at Berkeley and his co-workers have used enzymes to extract *DNA fragments* from muscle tissue attached to this pelt. They cloned the fragments in bacteria, compared them to zebra DNA, and found (as expected) that the genes showed a close evolutionary relationship. They have also succeeded in extracting and replicating DNA from a century-old bison pelt and from millennia-old mammoths preserved in the arctic permafrost. This success is a far cry from cloning a whole cell or organism—cloning one gene leaves about 100,000 uncloned,

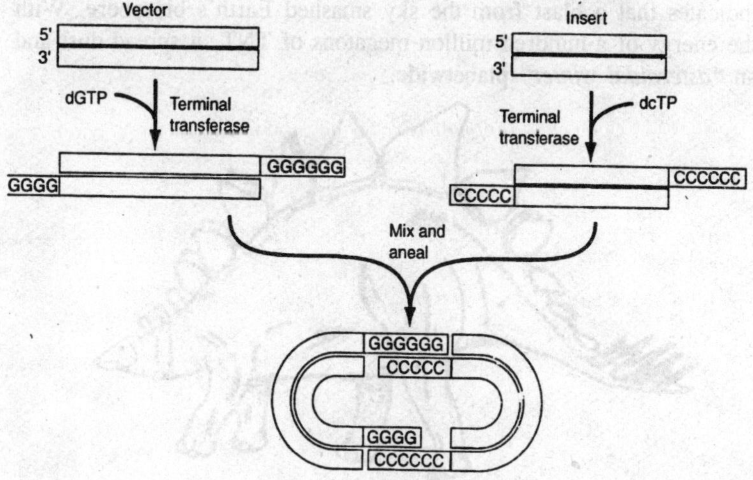

Fig. 8.4. DNA cloning.

and cloning every gene still doesn't repair a single cell—but it does show that the hereditary material of these species still survives.

Machines that compare several damaged copies of a DNA molecule will be able to reconstruct an undamaged original—and the billions of cells in a dried skin contain *billions* of copies. From these, we will be able to reconstruct undamaged DNA, and around the DNA we will be able to construct undamaged cells of whatever type we desire. Some insect species pass through winter as egg cells, to be revived by the warmth of spring. These "*extinct*" species will pass through the twentieth century as skin and muscle cells, to be converted into fertile eggs and revived by cell repair machines.

Dr. Barbara Durrant, a *reproductive physiologist* at the San Diego Zoo, is preserving tissue samples from endangered species in a cryogenic freezer. The payoff may be greater than most people now expect. Preserving just tissue samples doesn't preserve the life of an animal or an ecosystem, but it does preserve the genetic heritage of the sampled species. We would be reckless if we failed to take out this insurance policy against the permanent loss of species. The prospect of cell repair machines thus affects our choices today.

Extinction is not a new problem. About 65 million years ago, most then-existing species vanished, including all species of *dinosaur*. In Earth's book of stone, the story of the dinosaurs ends on a page consisting of a thin layer of clay. The clay is rich in iridium, an element common in asteroids and comets. The best current theory indicates that a blast from the sky smashed Earth's biosphere. With the energy of a hundred million megatons of TNT, it spread dust and an "*asteroidal winter*" planetwide.

Fig. 8.5. Stegosaurus (dinosaur).

In the eons since living cells first banded together to form worms, Earth has suffered five great extinctions. Only 34 million years ago—some 30 million years after the dinosaurs died—a layer of glassy beads settled to the seafloor. Above that layer the fossils of many species vanish. These beads froze from the molten splash of an impact.

Meteor Crater, in Arizona, bears witness to a smaller, more recent blast equaling that of a four-megaton bomb. As recently as June 30, 1908, a ball of fire split the Siberian sky and blasted the forest flat across an area a hundred kilometers wide.

As people have long suspected, the dinosaurs died because they were stupid. Not that they were too stupid to feed, walk, or guard their eggs—they did survive for 140 million years—they were merely too stupid to build telescopes able to detect asteroids and spacecraft able to deflect them from collision with Earth. Space has more rocks to throw at us, but we are showing signs of adequate intelligence to deal with them. When nanotechnology and auto mated engineering give us a more capable space technology, we will find it easy to track and deflect asteroids; in fact, we could do it with technology available today. We can both heal Earth and protect it.

Long Life and Population Pressure

People commonly seek long, healthy lives, yet the prospect of a dramatic success is unsettling. Might greater longevity harm the quality of life? How will the prospect of long life affect our immediate problems? Though most effects cannot be foreseen, some can.

For example, as cell repair machines extend life, they will increase population. If all else were equal, more people would mean greater crowding, pollution, and scarcity—but all else will not be equal: the very advances in *automated engineering* and *nanotechnology* that will bring cell repair machines will also help us heal the Earth, protect it, and live more lightly upon it. We will be able to produce our necessities and luxuries without polluting our air, land, or water. We will be able to get resources and make things without scarring the landscape with mines or cluttering it with factories. With efficient assemblers making durable products, we will produce things of greater value with less waste. More people will be able to live on Earth, yet do less harm to it—or to one another, if we somehow manage to use our new abilities for good ends.

If one were to see the night sky as a black wall and expect the technology race to screech to a polite halt, then it would be natural to fear that long-lived people would be a burden on the "*poor, crowded world of our children.*" This fear stems from the illusion that life is a zero-sum game, that having more people always means slicing a small pie thinner. But when we become able to repair cells, we will also be able to build replicating assemblers and excellent spacecraft. Our "poor" descendants will share a world the size of the solar system, with matter, energy, and potential living space dwarfing our entire planet.

This will open room enough for an era of growth and prosperity far beyond any precedent. Yet the solar system itself is finite, and the stars are distant. On Earth, even the cleanest assembler-based industries will produce waste heat. Concern about population and re sources will remain important because the exponential growth of replicators (such as people) can eventually overrun any finite re source base.

But does this mean that we should sacrifice lives to delay the crunch? A few people may volunteer themselves, but they will do little good. In truth, life extension will have little effect on the basic problem: exponential growth will remain exponential whether people die young or live indefinitely. A martyr, by dying early, could delay the crisis by a fraction of a second—but a halfway dedicated person could help more by joining a movement of long-lived people working to solve this long-range problem. After all, many people have ignored the limits to growth on Earth. Who but the long-lived will prepare for the firmer but more distant limits to growth in the world beyond Earth? Those concerned with long-term limits will serve humanity best by staying alive, to keep their concern alive.

Long life also raises the threat of *cultural stagnation*. If this were an inevitable problem of long life, it is unclear what one could do about it—machine-gun the old for holding firm opinions, perhaps? Fortunately, two factors will reduce the problem somewhat. First, in a world with an open frontier the young will be able to move out, build new' worlds, test new ideas, and then either persuade their elders to change or leave them behind. Second, people old in years will be young in body and brain. Aging slows both learning and thought, as it slows other physical processes; rejuvenation will speed them again. Since youthful muscles and sinews make young bodies more flexible, perhaps youthful brain tissues will keep minds some what more flexible, even when steeped in long years of wisdom.

Effects of Anticipation

Long life will not be the greatest of the future's problems. It might even help solve them. Consider its effect on people's willingness to *start wars*. Aging and death have made slaughter in combat more acceptable: As Homer had Sarpedon, hero of Troy, say, "O my friend, if we, leaving this war, could escape from age and death, I should not here be fighting in the van; but now, since many are the modes of death impending over us which no man can hope to shun, let us press on and give renown to other men, or win it for ourselves."

Yet if the hope of escaping age and death turns people from battle, will this be good? It might discourage small wars that could grow into a nuclear holocaust. But equally, it might weaken our resolve to defend ourselves from lifelong oppression—if we take no account of how much more life we have to defend. The reluctance of others to die for their ruler's power will help.

Expectations always shape actions. Our institutions and personal plans both reflect our expectation that all adults now living will die in mere decades. Consider how this belief inflames the urge to acquire, to ignore the future in pursuit of a fleeting pleasure. Consider how it blinds us to the future, and obscures the long-term benefits of cooperation. Erich Fromm writes: If the individual lived five hundred or one thousand years, this clash (between his interests and those of society) might not exist or at least might be considerably reduced. He then might live and harvest with joy what he sowed in sorrow; the suffering of one historical period which will bear fruit in the next one could bear fruit for him too." Whether or not most people will still live for the present is beside the point: the question is, might there be a sign change for the better?

The expectation of living a long life in a better future may well make some political diseases less deadly. Human conflicts are far too deep and strong to be uprooted by any simple change, yet the prospect of vast wealth tomorrow may at least *lessen* the urge to fight over crumbs today. The problem of conflict is great, and we need all the help we can get.

The prospect of *personal deterioration* and death has always made thoughts of the future less pleasant. Visions of pollution, poverty, and nuclear annihilation have recently made thoughts of the future almost too gruesome to bear. Yet with at least a hope of a better future and time to enjoy it, we may look forward more willingly. Looking forward, we will see more. Having a personal stake, we will care more. Greater hope and foresight will benefit both the present and posterity; they will even better our odds of survival. Lengthened lives will mean more people, but without greatly worsening tomorrow's population problem. The expectation of longer lives in a better world will bring real benefits, by encouraging people to give more thought to the future. Overall, long life and its anticipation seem good for society, just as shortening life spans to thirty would be bad. Many people want long, healthy lives for them selves. What are the prospects for the present generation?

Progress in Life Extension

Hear Gilgamesh, King of Uruk: "I have looked over the wall and I see the bodies floating on the river, and that will be my lot also. Indeed I know it is so, for who ever is tallest among men cannot reach the heavens, and the greatest cannot encompass the earth."

Four millennia have passed since Sumerian scribes marked clay tablets to record *The Epic of Gilgamesh*, and times have changed. Men no taller than average have now reached the heavens and circled the Earth. We of the Space Age, the Biotechnology Age, the Age of Breakthroughs—need we still despair before the barrier of years? Or will we learn the art of life extension soon enough to save ourselves and those we love from dissolution?

The pace of biomedical advance holds tantalizing promise. The major diseases of age—heart disease, stroke, and cancer—have be gun to yield to treatment. Studies of aging mechanisms have begun to bear fruit, and researchers have extended animals' life spans. As knowledge builds on knowledge and tools lead to new tools, advances seem sure to accelerate. Even without cell repair machines, we have reason to expect major progress toward slowing and partially reversing aging.

Although people of all ages will benefit from these advances, the young will benefit more. Those surviving long enough will reach a time when aging becomes fully reversible: at the latest, the time of advanced cell repair machines. Then, if not sooner, people will grow healthier as they grow older, improving like wine instead of spoiling like milk. They, will, if they choose, regain excellent health and live a long, long time.

In that time, with its replicators and cheap spaceflight, people will have both long lives and room and resources enough to enjoy them. A question that may roll bitterly off the tongue is: "When? ...Which will be the last generation to age and die, and which the first to win through?" Many people now share the quiet expectation that aging will someday be conquered. But are those now alive doomed by a fluke of premature birth? The answer will prove both clear and startling.

The obvious path to long life involves living long enough to be rejuvenated by cell repair machines. Advances in biochemistry and molecular technology will extend life, and in the time won they will extend it yet more. At first we will use drugs, diet, and exercise to extend healthy life. Within several decades, advances in nanotechnology will likely bring early cell repair machines—and with the aid of automated engineering, early machines may promptly be followed by advanced machines. Dates must remain mere guesses, but a guess will serve better than a simple question mark.

Imagine someone who is now thirty years old. In another thirty years, biotechnology will have advanced greatly, yet that thirty-year-old will be only sixty. Statistical tables which assume no advances in medicine say that a thirty-year-old U.S. citizen can now expect to live almost fifty more years—that is, well into the 2030s. Fairly routine advances (of sorts demonstrated in animals) seem likely to add years, perhaps decades, to life by 2030. The mere beginnings of cell repair technology might extend life by several decades. In short, the medicine of 2010, 2020, and 2030 seems likely to extend our thirty-year- old's life into the 20405 or 2050s. By then, if not before, medical advances may permit actual rejuvenation. Thus, those under thirty (and perhaps those substantially older) can look forward—at least tentatively—to medicine's overtaking their aging process and delivering them safely to an era of cell repair, vigor, and indefinite life-span.

If this were the whole story, then the division between the last on the road to early death and the first on the road to long life would be perhaps the ultimate gap between generations. What is more, a gnawing uncertainty about one's own fate would give reason to push the whole matter into the subconscious dungeon of disturbing speculations.

But is this really our situation? There seems to be another way to save lives, one based on cell repair machines, yet applicable today. As described already, *repair machines* will be able to heal tissue so long as its essential structure is preserved. A tissue's ability to metabolize and to repair itself becomes unimportant; the discussion of biostasis illustrated this. *Biostasis*, as described, will use molecular devices to stop function and preserve structure by cross-linking the cell's molecular machines to one another. *Nanomachines* will reverse biostasis by repairing molecular damage, removing cross-links, and helping cells (and hence tissues, organs, and the whole body) return to normal function. Reaching an era with advanced cell repair machines seems the key to long life and health, because almost all physical problems will then be curable. One might manage to arrive in that era by remaining alive and active through all the years between now and then—but this is merely the most obvious way, the way that requires a minimum of foresight. Patients today often suffer a collapse of heart function while the brain structures that embody memory and personality remain intact. In such cases, might not today's medical technology be able to stop biological processes in a way that tomorrow's medical technology will be able to reverse? If so, then most deaths are now prematurely diagnosed, and needless.

9

Nanotech Hobbies

Western society has long been a culture of abundance, its people spending a great deal of time, money, and effort in "the pursuit of happiness." People go dancing, see movies, and visit amusement parks. They build model railroads, collects stamps, garden, and pursue crafts. Others explore more esoteric hobbies, like war-gaming, bungee jumping, and dressing as medieval lords and ladies.

Nanotechnology will affect *hobbies* as dramatically as it affects everything else. Far and away the biggest impact, however, will be from the unprecedented abundance that nanotechnology makes possible, which will remove the need, for most people, to spend much time earning a living. People will have much more time to pursue hobbies—most of their lives if they so desire. Nanotechnology will also make possible more extravagant, gigantic, and imposing hobbies. Here are some examples of the possibilities.

Model Railroads

Nanotechnology is the most extreme form of miniaturization using ordinary matter, and because miniaturization is at the heart of model railroading, one might expect the two to be a natural match. A rotary engine can be at least as small as 30 nm in diameter, as demonstrated by bacteria that use such engines to move. Thus one could build a model railroad where the rails are 50 nm apart (a scale ration of over a billion to one.) Because a single engine or car would be near a light wave in length, a train would be much too small to see directly. On the other hand, one could build a working replica—small but visible—of the entire railroad network.

Actually, model railroading is usually not concerned so much with making things small as with making them realistic at the selected scale. At any current model railroad scale, nanotechnology could realistically render the shading on a splinter inside the letter "e" carved into the back of a wooden seat in a passenger car. Every detail that could be visible in a model railroad could be created using nanotechnological tools, although with the danger that tiny pieces—such as HO-scale playing cards—might blow away.

Most model railroads include *human figurines*. Because nanotechnology will be able to produce motors down to a millionth of an inch across, and hundred-million MIPS computers that will fit in a Z-scale skull, figurines that move and act realistically could be built for every model railroad scale. Some model railroaders may be reluctant to have nanotechnology provide everything. After all, a great deal of the satisfaction in making layouts comes from doing the work oneself. There are two responses to such reactions. The first is that much of that work requires delicate tools, and nanotechnology can provide the tools to create finer details oneself. The second response is that some people who might want to start model railroading find the initial setup too daunting; making the first steps easier might entice more people into the hobby.

Philately

Many hobbies involve collecting things and stamps and coins are classic items to collect. How will nanotechnology be incorporated into stamps, and how will this effect stamp collectors? One might reasonably argue that with massive computer networks linking everybody, postal mailing will disappear. That is possible, but often technology displaces old forms without completely eliminating them. Western Union, for example, still sends telegrams. Assuming postal mail continues to exist, how might nanotechnology affect the stamps? There are many possibilities. A networking model railroad mentioned above, could fit on a postage stamp. It would certainly be an original commemorative tissue. Indeed, working models and detailed scaled miniatures of anything could be produced cheaply for inclusion in stamps. Stamps themselves might change so that each included a nanocomputer, with unique serial numbers to reduce copying. How valuable would a Lincoln stamp with the serial number 05.15.1865 be.

Numismatics

Originally, coins had the value of the metal they contained. A coin was simply a way of certifying that a particular hunk of metal held the proper weight of gold or silver. Through many intermediate steps, we have come to have fiat money, which governments say have value, and which does have value when others accept the money in exchange. Thus, a fifty-dollar bill does not carry fifty dollars worth of ink and paper, but is rather a symbol for the wealth of fifty dollars. *Nanotechnology* will make possible atomically perfect counterfeits, so using money that is merely a physical symbol will no longer be effective. Instead, every coin could be given a unique serial number—but then a counterfeiter could pass different copies at different stores, unless a central data registry recorded who held what coins and bills. In that case, people might as well use the data registry and forego the actual coins and bills.

If this happens, the supply of new coins will end, but the demand to hold collected coins will likely stay at least the same, and may well grow dramatically when people have more time and wealth for hobbies. A possible answer to the counterfeiting problem is to return to coins that have value based on their content. The only way to perfectly counterfeit a one-ounce gold coin is to use an ounce of gold. This will be popular with some people, since not everyone wants the details of their finances kept in a central computer. If this sort of currency became predominant, it would turn the fiat money of today into interesting curios. In the case of both stamps and coins, authentic pre-nano-technological specimens will be rare relative to the wealth that follows, and may be prized as tokens embodying those older times.

Copying Issues

Nanotechnology will allow one to make perfect duplicates with atomic precision. (To make convincing copies, forgers must not only put the same atoms in the same places, they must use the same isotopes as well. Otherwise, a forgery could be recognized if it had, for example, the wrong amount of carbon 14.) How will collectors know they have an original? The easiest way to distinguish originals from copies is to put a mark on the original. Today, invisible serial numbers are etched into diamonds. This approach will foil those who create copies from detailed photographs of an original, but it will not work if the forger can use the original itself in making a copy. One approach would be to create a registry to authenticate that an original is an original. For collectibles that predated nanotechnology, this would be persuasive—if you could verify that the authentication record predated nanotechnology. Anyone with a valuable collection today will likely want to prepare a complete inventory and, if possible, register that inventory.

On the other hand, not everyone can have an original. By creating reproductions of significant items, more people will be able to become collectors. For example, an indistinguishable copy would be good enough for many who collect comic books. (A commercially legitimate reproduction would likely have "*copy*" unobtrusively written all over it, perhaps in lettering eight atoms tall, rather than be truly indistinguishable, but this will be unnoticeable for anything besides checking authenticity). A final difficulty—if one were to take an authentic Roman coin, disassemble it, and then reassemble two atomically precise copies, each with 50 percent of the atoms from the original, all in their respective original places, could one really say that neither coin was the original?

Gardening

There are those who like to garden, and others who think it is just an outdoor chore, but that's what makes a hobby a hobby. One of the biggest problems for those in the first group seems to be pests. Every plant is a living thing, and there are always other living things that would like to eat it. Although it may be an incredibly complex task, it may be possible to build *ecosystem protectors*. Such devices would not need to replicate, but they could be designed to kill specific weeds or animals as effectively as any predator. Ecosystem protectors could guard against invading species such as kudzu, fire ants, and killer bees in the United States, or rabbits in Australia. The ecosystem

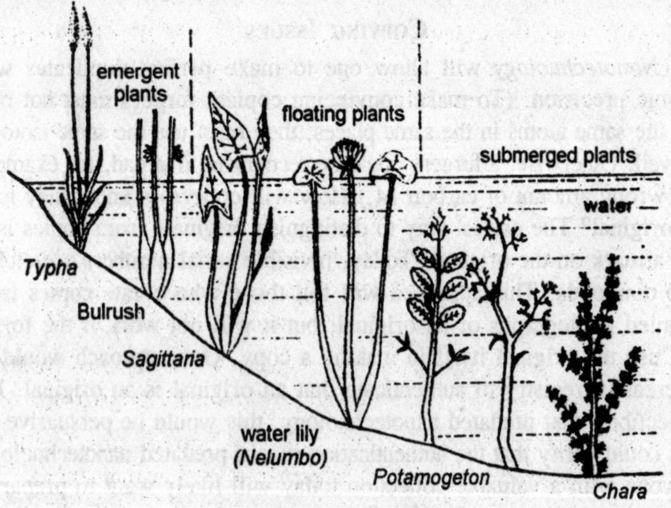

Fig. 9.1. Ecosystem.

a gardener is trying to protect is his or her garden. Nanotechnology could provide "*garden protectors*," to effectively kill off the unwanted pests invading one's tiny plot, without using chemicals could be used to create new, hardy strains of plants that look and smell exactly like the original but are designed to thrive in utterly different climates. Thus, one could grow that looked like orchids next to a mountain cabin, and what looked like healthy blue fir on a sunny beach. The colouration of flowers, when they bloom, the fragrances, or any other feature of a plant could also be specified. One could even order personalized cuttings, so that each petal of a rose bore the cameo of someone in the family.

Any plant designed to delight or amuse must carry an extra burden that would put it at a competitive disadvantage in trying to survive in the wild. For added safety, however, these garden plants probably would not have any of the capabilities necessary to reproduce (although they could present asexual flowers.) There is one plant a gardener could grow that seems almost mythic, a wish-for-it tree. Once the tree had grown large enough, one could walk up and ask for any object—a cup, a bicycle, a computer, a bucket of utility fog. Companion glasses, or any other small object that could be made of available atoms and that the tree knew how to build or could design. Then, a bud would form on one of the branches. It would grow and grow, and eventually one could pick the desired object off the tree.

Home Crafts

In a *nanotechnology culture*, *home crafts* will not be just an inexpensive way to build a coffee table or weave a nice table linen, but a means for people to express themselves in a tangible object and give something of their labour and human spirit to one another. Every object a person makes will have sentimental value.

Many people who like to build things have workshops in the garage because it is the easiest and least disruptive room in a house to convert into a shop. In a nanotechnology home, however, one could easily transform any other room in a workshop. On demand and with a little notice, the tools and work pieces could be put away, and the shop would turn into the old room again. The shop could have the finest hand tools possible with the sharpest cutting edge atoms can form. The handles could change their shape to fit the hands perfectly and to massage away muscle pains. *Molecular intelligence* on the work end would make smart tools that adjusted to the work piece, or that modified it at the atomic level, like varnishes, but with vastly finer control.

One could also incorporate nanotechnology in the work piece, including, say, a video game in an ornately carved wooden board, or making a stool that was self-repairing.

Needle point seems to be *immune to changing times*. It is obviously not a cost-effective way to decorate cloth, so it is clearly serving other needs and purposes. One purpose is that needlepoint is much more personalizable than manufactured cloth; most gift samplers include spaces for names and dates for births or weddings. In a culture of abundance, however, we will be able to provide cloth personalized to each individual thread no more expensively than bulk cloth is produced now.

The second function of needle point is the *sentimental value* mentioned above, and the final benefit is the pleasant, calming effect it has on those who do it. For this hobby, nanotechnology could help by removing some of the petty annoyances, by offering self-threading needles for example. A piece of cloth could have a pattern printed directly on it that faded away at each stitch, or could sound a gentle alarm if one mis-stitched a thread. For those who want to experiment with the final product, there would be active threads that move on their own, creating interesting visual effects in the final design.

Nanotechnology will make it incredibly easy to build specified objects, so designing furniture, pottery, textiles, and so forth might become a popular pastime practiced by those with artistic inclination. Final products being fabricated automatically.

Blood Sports

Not everyone spends their hobby hours in quiet seclusion; the *adrenalin* rush in widely popular. Witness bungee jumping. *Molecular engineering* could provide super bungee cords of varying thickness and much stronger material that would allow significantly higher jumps. Other approaches could work, and all would protect people who decided to jump out of airplanes without a parachute. Most daredevil activities are similar in that they have in common experiences of speed and acceleration. Future wealth will provide the opportunity for a great deal of both, and nanotechnology will provide the means to help people survive the experience.

Henson has suggested that hand-to-hand combat, using *medieval weapons*, could provide a thrill for those seeking historical entertainment in the future. This activity could be quite realistic, down to producing wounds that today would kill a person. One difficulty is that the major damage to the brain may not be repairable even with nanotechnology.

Those portions of the mind where information has been lost might not return on reconstruction. Truly severe head wounds could be fatal. Such games may well draw the more extreme thrill seekers, for whom nanotechnology may seem to have taken all the "*fun*" from life.

This brings up a moral issue: Should people engage in *blood sports* when their direct threat to life becomes small? This is not the question, "Should blood sports remain illegal?" With nanotech healers standing nearby, blood sports could be made less dangerous than sports car racing is today. If "*play*" is a result of the informed choices of consenting adults, many would consider outlawing that choice overly authoritarian, an infringement of rights. The moral issue arises when we consider that repeated exposure to brutality may lead to an acceptance of violent acts performed without mutual consent, If, as it seems future technology will allow us to much more easily translate our thoughts into reality, we may want cultivate finer thoughts than blood sports would lead us to contemplate.

Thrill Rides

The classic thrill ride is the roller coaster, still growing in popularity, Higher performance bulk materials will make possible taller, faster roller coasters with more elaborate track layouts. With current materials, however, it is already possible to make roller coasters that pull "*snap turns*" sharp enough to kill people—current acceleration limits are based on rider safety, not the technological limits of roller coaster engineering. This, future roller coaster rides may not pull on people much more strongly than today. On the other hand, greater disposable wealth will mean that parks can afford much longer rides. Amusement parks have many other rides that spin, loop, lift and twirl. A towering Ferris wheel made primarily out of diamond would look spectacular on the fairway at sunset.

Amusement Park Attractions

Many people agree that the *Disney Corporation* offers the best amusement parks in the world. The Disney people insist, however, that the parks do not offer "*rides*," they offer "*attractions.*" Anyone who has been to a Disney park can sense what they mean by the difference. A Disney attraction aims to be a total experience. The parks are divided into tourist-friendly representations of places one might want to visit: a land of fairy tales, the American Old West, or the future. Each attraction has a theme that fits within its area, and each attraction looks like it belongs there. The line one waits in for an attraction continues the theme while offering diversion. The attraction itself is entertaining and seems more realistic than just a roller coaster. Increasingly, other amusement park developers are trying to copy Disney's approach.

With *nanotechnology*, very realistic attractions will be easy to create and affordable. Until then, Disney and other amusement park companies will continually push available technology in order to make the background, the characters, the action, the motion and the sounds and smells all seem more realistic. This will eventually make these attractions very believable, but it will not make the fantasies real. In other words, amusement parks will offer ever-improving virtual realities. Indeed, entertainment drives much of virtual-reality technology, because amusement parks can pay for simulations long before such equipment is affordable in the home.

Two of the most difficult effects to create are sensations of continuous extreme acceleration and weightlessness. This is because they can only be approximated by using real acceleration, which requires long distances if it is to be continuous. Thus, certain daredevil activities like jumping from airplanes cannot be easily co-opted by virtual reality—yet. Nanotechnology-based virtual reality will be able to simulate even these effects by actual continuous acceleration across the solar system, or holding one at rest away from any gravitational bodies. In any event, amusement rides will undoubtedly be a growing popular entertainment for the rest of this century and beyond. Whatever one's hobby, be it an old hobby in a new world, or a new hobby made possible by future changes, the abundance *nanotechnology* promises to provide will allow one to spend vastly more time pursuing that hobby. The machines can have dinner ready when you are done.

10

Human Prospect

One way to sum up nanotechnology is that it will make matter into software. *Nanotechnology* is, of course, hardware, but it has many more features in common with today's software than today's hardware. Fifty years ago, software was an arcane art practiced by an elite priesthood in rare, expensive, and impressive settings. Today, everyone uses it, be it on a computer, cell phone, or ATM. Given the *Moore's Law* expansion of processing power, what software can do is limited only by the imagination.

Software is incredibly intricate and complex. There is an enormous amount of technology that goes into software in and of itself: everything from the low-level details of a given computer's instruction set, to the operating system and its interfaces, to compilers and software development environments, to computer science with its mathematical analysis of algorithms. However, for any given application, there is an entire body of knowledge about the end user's field of interest that has to be added. For a word processor, it's knowledge of letters and words and fonts and formatting. For an engineering CAD system, it's knowledge of three-dimensional geometry measurement systems, drafting techniques, and the like. And on and on, for Web browsers, airline reservation systems, geneology database programs, the works.

In just the same way, nanotechnology will have a core of techniques required for any application. But for each end purpose, more and specialized knowledge will be required: biology for nanomedicine, aeronautics for flying cars, civil engineering for structures, and so forth. If you use a *computer* nowadays, what you actually interact with is the software. All the things you see: windows, pictures, text, menus, cursors, icons, and the like, are created by the software design, not the *physical computer*. The same physical computer, with different soft ware, could just as easily give you a completely different interface. Many applications do this: a flight simulator has you interacting with a completely different set of concepts than a recipe manager.

In the physical world, of course, the interfaces to an airplane and a kitchen are different, too. The cockpit controls look and act nothing like your range, sink, and refrigerator. But the feature of software that will carry over into nanotechnology is that the world you experience with software is completely made up. The only limit is the designer's imagination. With nanotechnology the world you experience—not just patterns of light on the screen but the whole physical world—could be entirely made up, as well. As with any new, glitzy technology, this seems likely to be overdone at first, but soon, let us hope, wiser counsels will prevail, and reasonable, moderate uses of technologies such as Utility Fog will be the norm.

A World of Your Own

For two million years or so, since those first chipped stones, advances in technology have redounded to the advantage of those who had it. Faced with an overwhelming and often hostile natural world, our ancestors needed all the help they could get. Technology extended their power and gave them more mastery over that world.

Since the invention of agriculture, real property and cities, however, there has been a subtle reversal in the trend, which has come to a head in the past few centuries. Once nature is essentially mastered, extending one's grasp can only encroach on the lives of other people. On the average, nobody can win. Life comes to resemble a zero-sum game in many ways. With an increasing population, it is worse, since each person's share shrinks.

A zero-sum game is a recipe for unhappiness and strife. We are built, psychologically, to enjoy doing good things for ourselves and for others. In a zero-sum game, you cannot do good for anyone without doing harm to someone else. The very aspect of the Prisoner's Dilemma that allows cooperation, and thus morality, to evolve, is that it is not a zero-sum game. Without the evolutionary pressure of the Prisoner's Dilemma dynamic, the basis of altruism and morality disappears and they begin to erode.

Humans always will—and should—compete, or they would not be human. But in a static world, empathy and kindness are repressed by the lack of opportunity. Add to this the sense of frustration created by a serious discrepancy between what one could do, in the absence of others, and what one is allowed to do. In reality, in order to gain in such a world, people will begin looking for ways to demonize others, so as to be able to hurt them with a clear conscience. Blame and injury breed deep-seated hatred and strife.

Such a world is an evil place, but it is one we are heading toward all too clearly, if the conventional vision of limited resources "just one Earth" holds sway. Technology both helps and hurts. When you sit in a traffic jam, annoyed at the other drivers, it seems the automobile is the root of the problem. But it is really people, as the pedestrian traffic jams in China's cities demonstrate. The auto is well matched to population densities outside of cities. But the auto, or any other technology that extends one's range, exacerbates the problem when that range overlaps other people's. People evolved to live in an uncrowned world and to associate on a more voluntary basis with fewer other people than they typically do now.

Technology helps by increasing total wealth. This makes the world less zero-sum in some ways. Increasing the total stock of value means you do not have to rob Peter to pay Paul.

One reason computers have become so successful over the past decades is that in a very real sense, they expand the world. With a computer, there are more things to do, more items of interest, and more people with whom you can associate or not as you please. We may be forced to rub elbows with strangers in the physical world, but there is plenty of room in the virtual one.

Nanotechnology can expand the world in many ways beyond the assisted sublimation of the computer. There really is plenty of room at the bottom. Acting through a virtual reality link, a two-millimeter *humanoid telerobot* would find a ten-acre field as vast and daunting as Henry Stanley and David Livingstone's Africa, and filled with stranger, more varied, and more dangerous creatures. Or you could live in literally limitless virtual worlds indistinguishable from physical reality to your senses. You could do more real, physical things in less space, including those *nanomachine* shops in your design workstation. And you could physically move away, living in remote areas on Earth or in outer space.

Elbow Room

A well-used basketball is covered with a layer of sweat, oils, dead skin cells, and living bacteria. This layer is thicker on the scale of the basket ball than the layer of life, the biosphere, is on the scale

Fig. 10.1. Homo sapiens.

of the Earth. This layer contains not only all intelligent life, but all known life of any kind, in the universe.

The *biosphere* seems huge and limitless to an individual human. For humankind as a whole, it is beginning to get a bit tight around the elbows. Edward O. Wilson, the eminent biologist, suggests that *Homo sapiens* may well already have exceeded the long-term carrying capacity of the biosphere in the late twentieth century. It is completely clear that we are well over the capacity of the environment to feed us, absent large-scale agriculture. And that means agriculture with irrigation and synthetic fertilizers—simple farming won't do it.

To get an idea of how precarious a position we have gotten ourselves into, simply consider what would happen if some new disease did to wheat and rice what the Dutch elm fungus did to elms in the United States since 1930. It would not really matter whether the disease were human-made or entirely natural. The human species is vulnerable because everything we are, need, and value is here in this vanishingly thin film of life on one planet. And yet we are increasing the load we place on the biosphere, both in resources per person and in the number of people.

How many people can the Earth hold? With nanotechnology, quite a few more than without it, just as agriculture itself increased the practicable population density. According to the US Census Bureau, the current population density of the United States is about eighty people per square mile. That is already too many for each person to have a ten-acre estate. The most densely populated area in the United States is Jersey City; New Jersey, with 13,044 people per square mile. If we were to populate the entire land surface of the Earth to Jersey City densities, it would hold about 750 billion people, or 125 times its current population. There would be no room for fields, farms, parks, forests, or anything; all food would have to be produced in the oceans (or synthesizers). Note that that population is only seven doubling times away from our current world population of 6 billion.

With *nanotechnology*, the prospect is less hellish. We can build mile-high towers covering only 1% of the land surface and house five times as many people while leaving 99% of the surface in a natural state. This would give us two and a fraction more doubling times. Rather than agriculture, *nanotech recycling* would stand in for the carrying capacity of the biosphere. At this point, though, living in one of these giant hives, you had depend on virtually the same technology that would be required to live in space, and visits to the real biosphere would be strictly regulated to prevent the swarming people from trampling it flat. And we are still in trouble at ten doubling times.

How long is a doubling time? Recent US growth rates have been about 1% per year which gives a doubling time of seventy years. Historical growth rates have been over 3%, for a doubling time of twenty-five years. It is generally believed that a rising level of affluence is correlated with declining growth rates, and that is all for the good. However, nanotechnology will most likely throw a monkey wrench in that machinery. *Nanomedicine* will very likely extend life times significantly, even indefinitely, and preserve youth, vigour, and childbearing ability for a much longer term. This means that people will have the ability to have a career, retire with a comfortable pension—and then have a family. What is more, the mortality column in the population tables will decline toward zero. Census Bureau projections for 2050 give, somewhat optimistically, a birth rate of 1.4%, a death rate of 0.98%, and a net immigration rate of 0 2 5%, for a growth rate of 0.67%, and a doubling time of over a century. But if the death rate goes to near zero, and childbearing age becomes twenty to one hundred instead of twenty to forty; we might easily get a 3% growth rate again. Worldwide Jersey City by 2200!

How about *restricting birthrates*, as China does? The picture is relatively stable with one kid per couple, but now there is a whole new reason to want to go to space: to have babies. The main problem with staying on Earth is that it is squarely against both our deepest instinctive drives, to stay alive and to have children. We have either got to stop making babies or start making land. The latter seems more preferable.

It is worth pointing out that land, in the form of O'Neill colonies, can be built in such a way as to accommodate extensions of Earth's biosphere. Population pressures on Earth put our values for babies and wildlife directly at odds with each other. The asteroids alone could provide the material for colonies with twenty thousand times the land area of Earth. No species need be crowded out of existence. Country or wilderness living can continue to be an option, although a full-fledged rotating, shielded colony will always be more expensive than a simple house in space.

If there is only one Earth, and everyone lives here, and everything that humans do is done here, retaining significant chunks of the biosphere as nature preserves will become harder and harder as time goes on. If we move into space, however, the Earth will ultimately be less than 1 % of 1% of the total livable area humans have available. It seems much more likely that a significant portion could then be set aside and a large proportion of living species saved from extinction.

In the long run, of course, we had better be living in other places than the Earth, or the next *dinosaur*-killer asteroid will drop the curtains on our little show. The chance of this happening in any given year, however, are about the same as your chances of both winning the lottery and being killed in an auto accident in the same year. People being what they are, would not count on any major projects being undertaken on the basis of that risk any time soon. Still, we will be safer as a species in the long run with our eggs in several baskets.

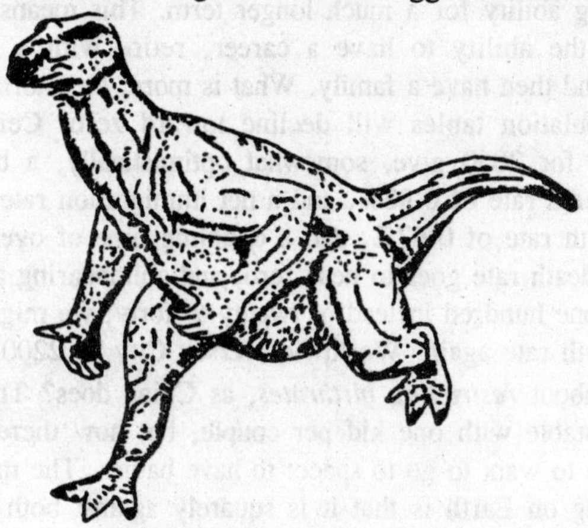

Fig. 10.2. Dinosaur (Iguanodon).

There are other dangers to worry about, though. It is always been the case that a major plague could wipe out a large fraction of humanity; We have grown complacent with modern *antibiotics*, and we may even have a right to be, as *nanomedicine* advances. However, the increasing sophistication of biotechnology makes it likely that quite nasty plagues could be created from scratch. This could happen either as a purely malicious act by a madman or terrorist or by semiaccident. Consider a scenario where delinquent youths release killer plagues for a thrill, believing that no major harm will be done because of advanced nanomedicine. This would correspond to today's "*script kiddies*" who write computer viruses. Then one of the plagues evolves to be just a bit trickier than the current defenses can handle. *Replicating nanomachine* infestations are also a possibility, though the biotech ones will be possible sooner. And good old-fashioned nuclear terrorism won't go away—nor will state-sponsored nuclear war.

Balance of Powers

Certainly it seems that nothing could have been more obvious to the people of the earlier twentieth century than the rapidity with which war was becoming impossible. And as certainly they did not see it. They did not see it until the atomic bombs burst in their fumbling hands. Yet the broad facts must have glared on any intelligent mind. All through the nineteenth and twentieth centuries the amount of energy that men were able to command was continually increasing. Applied to warfare that meant that the power to inflict a blow, the power to destroy, was continually increasing. There was no increase whatever in the power to escape.

There is a *natural balance* in the capabilities for destruction and transportation in any given level of technology. A man with a sword might slay a family unawares in his village but he is unlikely to best five other men with helmets and shields. A man on horseback can chase down peasants on foot, but others on horses of their own escape him. Chemical explosions confined in steel cylinders give the gun a great destructive range, but in the automobile they give people the ability to live out of gunshot from each other and still engage in daily commerce.

The most destructive technology of the twentieth century, the nuclear explosive, could similarly have extended the range of practical transportation. The *Nerva* and *Orion* rockets could have opened up the solar system, and nuclear power plants could have powered colonies on places like the Moon and Mars, and operations like O'Neill's lunar mass driver. This was not done, of course, and the impact of nuclear technology' has remained strongly biased toward the destructive. This seems to be typical of technology that remains largely in government hands, and the phenomenon has contributed to a perception of technology as more dangerous than good in the twentieth century. There is a pernicious feedback effect: the perception of dangerousness feeds the restriction, which in turn prevents the general public from gaining experience and finding beneficial uses for the technology. Thus most uses found are for military applications, feeding the perception of dangerousness.

It is not all a question of technology of course. The bounds of the Earth put us in a boxing ring that is large compared to horse or *automotive technology* but small compared to rocket and nuclear. With nanotechnology, it will shrink again. Biotech weapons are the first where a single person can carry something that would threaten a substantial fraction of humanity. They will not be the last. *Nanotechnology* gives us the ability to move outside the box. It makes possible single-family dwellings in space that are comfortable, self-sufficient, and hundreds of miles apart. A terrorist with a hydrogen bomb, facing such a spread-out population, would be reduced to the destructiveness of the man with a sword.

There are several implications. First is the obvious: given someone with such a weapon and the will to use it, he imposes a risk that would be cut by a factor of a million. That means that such an act would not have an near the political effect of Earth-based terrorism, so the motive to do it in the first place would evaporate. Life in space could be both free and safe. You could trust you neighbours even if you did not watch them constantly. A cluster of homes in a quintillion cubic miles, tiny on the scale of the solar system, could easily be a more tightly knit community than a twenty-mile square of city on the Earth's surface.

Back on Earth, we are all living in the same soup bowl. Any accidental or purposeful release, whether replicator or merely poison, pollutes the one atmosphere we all breathe. *Nanotech defenses* and counter measures are certainly feasible and, will prove to be effective; but once you are wearing a skinsuit and breathing filtered air anyway, the jump to a spacesuit does not seem all that great.

Finally is simply the question of the quality of life. People will usually take the choice of more room over less, all other things being equal. They will choose safety over danger, freedom over restrictions. They will have families. They will seek affluence over poverty.

A large part of affluence is energy. Current global energy consumption is about 2 kilowatts per person. This already has some effect on global climate. The effect can be reduced by moving away from *fossil fuels*, but we can increase energy use by only a factor of ten, at most one hundred, before the energy itself—even if its generation is absolutely clean—begins to affect the climate substantially. A large part of this headroom will be consumed merely in raising the existing population to Western levels of affluence. (Energy consumption in the United States is about 12 kilowatts per person.) Remember that it is rising affluence that demographers are counting on to *ameliorate* population growth rates.

If we were in space using solar power, and the population were one thousand times what it is now and each person were using one thousand times as much power as we do now, we had still be using less than a millionth of what the Sun puts out. The streets of space are paved with gold—golden sunlight.

Perhaps the most important part of the quality of life is that people need a challenge and an environment in which their exertions can affect their prospects, and those of their children, for the better. Humans need to face human-sized problems. Too hard and the result is continual failure and a broken spirit. Too easy, or no challenge at all, and the result is boredom, apathy, and ultimately, random acts of irrational destructiveness. With nanotechnology, Earth is just too small a sandbox. We must now step out and measure ourselves against the stars.

Morality and Centralization

In this book, it is tended to draw a picture of *nanotech applications*, as coming in a world where individual control and responsibility hold away, and the benefits are aimed at individuals. The twentieth century saw a wide spread trend toward another extreme, where responsibility moves from individuals to formal organizations. It might be reasonably asked whether the trend will actually stop, or whether we will ultimately wind up essentially pets of the government, with the right to choose which flavour of ice cream we want but not to do anything that makes a significant difference in the world.

One thing that is happening is that there are more people, but less work. As manufacturing and other systems become more capable and efficient, the jobs are spread more thinly, making each person responsible for less and less. Such specialization makes people more efficient but has a tendency to shift responsibility, power, and rewards, both psychic and material, from the individual to the organization. Ironically it was Karl Marx who was first noted for discussing this phenomenon, which he called alienation.

Two other obvious pressures are behind the trend. First is simply that organizations will take power when they can. This is a purely evolutionary dynamic, since organizations that take power will tend to displace ones that don't. Second is that people will use organizations to shield themselves from responsibility wherever possible. In the long run, of course, when you avoid the responsibility for something, you give up the corresponding power as well. Thus things that used to be in the purview of personal common sense are now centrally con trolled and regulated.

It is not as if any common sense were directing the central regulation The US government regulates the number of ounces of water can use to flush your toilet, but sells irrigation water for less than penny a ton to rice growers in semi desert areas. The FDA causes the death of thousands of citizens every year by refusing to certify drugs that are widely in use in Canada and Europe, in the name of safety. Meanwhile, federal agents arrest people for using marijuana for medical purposes (to relieve glaucoma, for example), although such uses are explicitly legal there. Everyone has his favourite example of the foolishness of centralized control. The Allies wanted to impede the German war machines by constricting the supply of ball bearings. So they sent a bomber raid to demolish the headquarters of the nationalized German ball bearing industry. Ball bearing production tripled.

In some areas, such as economics, it's come to be understood that central control is a disaster. Economies work only because of the widespread information gathering and decision making done by everyone all the time as buyers and sellers. In 1900 serious, thoughtful people could believe, and many did, that rational, scientific control of the economy could be more efficient and less wasteful than the helter-skelter hodgepodge of the marketplace. But after some large-scale experiments, such as the Soviet Union, a deeper understanding was gained.

Nanotechnology will bring the issues of personal responsibility to a head. For example, should people be allowed to make drugs and weapons in their home synthesizers? How about two-gallon-per-flush toilets? Medical equipment for use in emergencies? The crucial issue is that nanotech and AI could make it possible for an extension of the state to be watching every person every minute and controlling every thing they do.

Humans were evolved to be *semiautonomous*. We prefer to live in small groups but are, in our natural state, quite capable of existing as individuals or family units. Large groups of hunter/gatherers depleted game and forage stocks and were not favoured. With agriculture, more substantial groupings evolved. The largest concentrations of people, even in antiquity, were associated with considerable technology: the aqueduct and sewer systems of Roman cities, for example. With the coming of the industrial revolution, however, the "*society as machine, people as parts*" model really took hold.

It might reasonably be projected that nanotechnology, as an extension of the industrial revolution, might simply continue the trend. But that is not necessarily the case. Today's economy is highly interdependent: a complex product may have parts or materials from a hundred countries. Nanotechnology can reverse this completely. Your countertop synthesizer will have more moving parts than there are in all the machines on Earth today. The trend to complexity can continue, but at the same time we have the opportunity; if we desire it, for a level of autonomy, independence, and freedom, equivalent to that of the hunter/gatherer.

New Frontiers

Nanotechnology offers frontiers in many directions that were not available before. The solution to crowding and resource depletion is clearly to move into outer space. If you look at it from a broad perspective, space is the real universe and the Earth is a microscopically small bubble in a vast sea of room and resources. It is ridiculous to stay here when we have so clearly outgrown our original niche as a bipedal ape.

Space sounds farfetched because the practical difficulties are substantial. A *shuttle flight* requires thousands of technician-years of maintenance for preparation. How can something like that ever be economical? Perhaps a final comparison with computers will help. When you open a document in a word processor, there is generally at least a one-second pause before the document appears on your screen. In that second, a modern PC will have done a billion operations as complex as multiplying ten-digit numbers. Just fifty years ago, a clerk doing just the same bookkeeping or arithmetic would have taken at least a minute for each one. Thus the one-second pause of your word processor represents nearly two thousand solid years of clerical work.

As nanotechnology matures, the same ability to throw what would have been enormous efforts at the most trivial problems will come to the physical world as we have in the software world now. Living in space is dangerous and prohibitively expensive with current technology; it will be cheap, easy, and safe with advanced nanotechnology. We really can simply leave our problems behind.

Uploading

Uploading means copying your brain and as much of the rest of your nervous system and bodily reactions as necessary, in software form, into a computer or other substrate that gives you more speed, processing power, memory, longevity; and room to grow than the original biological equipment. Some people might want to exist as individual robots, while others would prefer to live in the same huge computer, with much higher communication rates between them. This

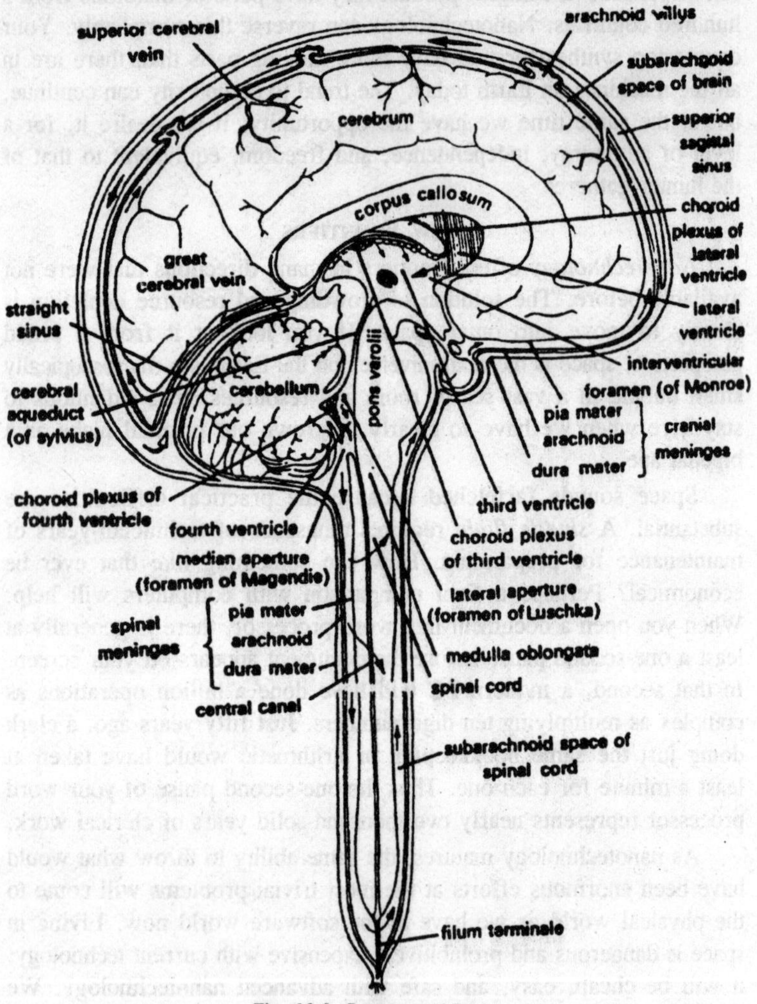

Fig. 10.3. Structure of Brain.

is obviously a choice not everyone has to make the same way. We can expect to find far-flung scatterings of rugged individualists and centralized concentrations of the gregarious. For people living in the more crowded of accommodations, it would make sense to live primarily through virtual reality. Interactions with the physical world would be by virtual reality links to android robots, or for extended excursions, mounting the brain in one.

The physical human brain is slow and fragile, and has idiosyncratic dietary requirements. Better, perhaps, to reimplement it in a software substrate. This would have some significant benefits. You could download yourself into a robot instead of having to use a net work connection or go to the bother and danger of having it physically carry your brain around. You could further avoid danger by making backup copies of yourself, in more than one place. You could avoid tedious journeys by transmitting yourself as data from one robot to another. When not using a physical body, you could run, connected to virtual reality worlds, on powerful stationary processors thousands of times faster than physical-world subjective time rates.

On the other hand, it will probably take a while to convince you that once you get your brain copied off into software, it is really still you. After all, we have pointed out that atom-for-atom teleportation by analysis and reconstruction needs data rates too high for even likely nanotech networks. The difference lies in the fact that the mechanism of the brain that gives rise to thought, consciousness, memory, and so forth is almost certainly at a higher level than the molecular. Most of what goes on in brain cells is the same stuff that goes on in muscle, skin, and fat cells: DNA and hormones regulating the activities of ribosomes and metabolic pathways, and mitochondria providing ATP to power those and other activities.

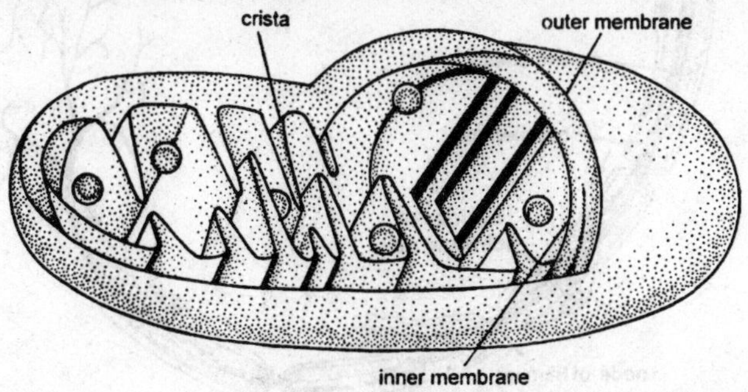

Fig. 10.4. Mitochondria.

It is very likely that the level of activity that supports our consciousness is the firing, synaptic transfer, and internal processing of the electrical signals in the neurons. It is certainly true that they are crucial to all the brain, sensory; and motor functions that have been studied. Just as certainly, other inputs and processes are involved, but they are not mysterious; merely as yet incompletely understood. It only reasonable to assume that the same higher-level processes could e supported by a different set of fuel-burning, power-generating, and construction machines.

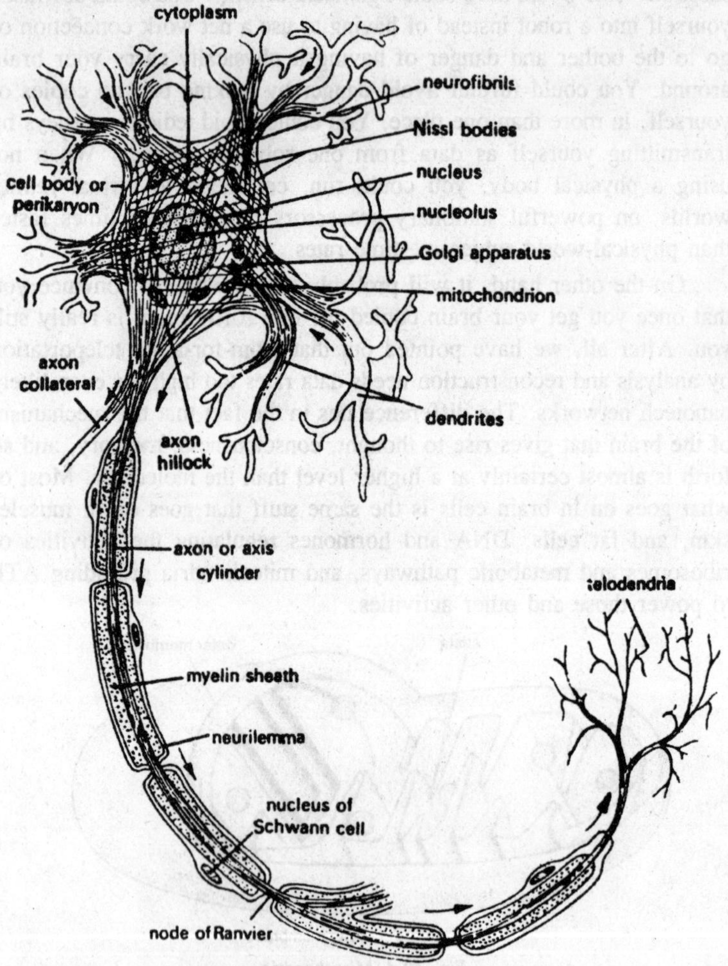

Fig. 10.5. Neurons.

This is where most people's intuitions about the process break down. "Okay," they say, "I'll grudgingly grant that you could build a machine that would act like a brain, and maybe even act like my brain, but it would not really be me. If you took my brain apart to create a software version, you had have killed me and created a new robot that imitated me in a ghoulish fashion."

One way to get around the intuition is the scenario by Hans Moravec in his book *Mind Children*. He imagines that you can be conscious while the conversion is being done, and that it is incremental, a small clump of brain tissue at a time. The process is also reversible, so that you can experience being in the computer form, return to your were brain, and remember the entire episode. If, when this happens, you experience continuity of consciousness and your memories are consistent, it becomes harder to argue that it was not really you. The crucial joint is that, having experienced it, you would believe it was really you.

Not everyone's intuition is the same, of course. Many people believe in *reincarnation*, and a widespread audience was capable of the suspension of disbelief to make *My Mother the Car* at least comprehensible. Almost certainly there will be people at the opposite extreme who will refuse uploading no matter what the evidence. As long as it remains a matter of individual choice, this does not seem to be a problem.

We undergo discontinuities of consciousness every night in the ordinary world. In normal experience, the continuity of the physical body establishes identity but we don't have any alternative to compare it with in formulating our expectations. Nanotechnology will make Moravec's scenario possible. My own expectation is that if it does work as envisioned, the intuition will change after examples become numerous and everybody knows one personally.

Uploading offers another way into a bigger world. As wide-open as the physical possibilities are with nanotechnology, they are wider still uploaded. The current-day philosopher asks, "What is it like to be a bat?" but the upload could *know*. We could have new senses, not merely mapped onto our current set, and new forms of intuition, may be even new emotions, more appropriate to the world we live in.

You must not think of such a world as over complex and confusing. It would be to us, but so would our world be confusing to *Homo erectus*. In fact, our descendants (and with a little luck, maybe even ourselves) will be more naturally comfortable and understand their environment more intuitively than we do ours today. That is because we have jacked up the complexity of our current world, but not the equipment we use to understand it; they will be able to do both.

Where does personal responsibility and independence go when people are programs running on the same ultra-mega computer? Perhaps surprisingly, the range of options is the same, or perhaps even wider, than in the physical world. Let's consider a few cases, as widely scattered signposts to the vast terrain of possibilities.

There could be the equivalent of a processor per person, with communications channels between them, and one or more complex environment simulations for them to interact in. This would correspond to people with separate brains in the real world. This level of integration would interact well with real humans and people running on physically separate robot processors. The assumption here is that your thoughts are entirely yours, and that you could own a part of the physical world or simulated environment over which you would exert more or less exclusive control.

In a software world, it will be possible to create the equivalent of germs, fleas, lice, and ticks—the descendants of computer viruses—simply by thinking about them. The temptation will be great for the community to want to control your thoughts in fear of such things, even though people who did that would be as rare as people who deliberately spread disease today.

This is a significant concern because lowering the firewalls between people will have so many advantages in other ways. Exactly the same kinds of thing happened to people when they began living in cities: disease was a scourge, and epidemics like the *Black Death* could wipe out a third of the population. Yet people crowded into cities because it greatly facilitated communication and trade, the building of common infrastructure, and other economies of scale. And yes, there were plagues; but the advantages (usually) outweighed them. Indeed, living in such cities clearly made people stronger and more effective in the long run.

In the physical world, technology has helped finesse the issue, with transportation, sanitation, medicine, and so forth. *Nanotechnology* can carry that further, for example, with skinsuits acting as biological firewalls but allowing direct personal contact. In the software world, the choices are harder. Uploading will allow things like direct transfer of thoughts and emotions, joint experience, and many modes of interaction as yet unthought-of. It will also allow not only direct monitoring of people's thoughts, but legislated changes in the structure of their minds. Given the track record of bureaucracies in the real world, the clear and present danger is that communities of uploads would quickly evolve into soulless monstrosities.

It should be noted that the ability to modify people is not limited to the uploads—it will happen all too soon in the physical world with brain-altering *nanomedicine*, *genetic engineering*, and the like. Indeed, the level of technology, both physical manipulation capabilities and the scientific understanding of what to manipulate, that will enable uploading in the first place is almost certainly capable of restructuring biological brains instead.

Luckily soulless monstrosities wound not win in the long run. They just can't seem to play nice with other soulless monstrosities. Evolution could have taken us that war, like ants, but did not. There is too much value in the adaptable flexibility of the semiautonomous intelligence that we are. One of the challenges awaiting us as we move forward is to understand this well enough to avoid some unfortunate experiments.

With a properly defined Bill of Mental Rights, however, an upload community could be a truly marvelous place. It would be like the concentration of talent of a Hollywood or a Silicon Valley—centers of great creativity and an enormous value to humanity as a whole.

Not only does not everyone have to make the same choice, but each of us will have all the options, and can be biological humans, physically autonomous robots, and members of upload communities, serially or even at the same time. It will be trivial to copy yourself into different forms or many individuals of the same form, and not so trivial but possible to merge them back again. You can know the tight brotherhood of an upload community where you can literally feel your neighbour's emotions, the adventure of a millimetric ranger keeping the insects of Earth at bay, the thrill of a thousand-ton robot exploring the moons of the outer planets, and relax as a standard-issue human playing golf on a terraformed Mars.

To Thine Own Self be True

Are you the same person you were ten, twenty, forty years ago? People grow and change. Literature in which the characters don't develop is considered flat and second rate. We are *autogenous creatures*, creating ourselves as we go along. Nanotechnology will give us the ability to improve ourselves, a noble pursuit, but one fraught with enigma and danger.

Living in a biological brain puts bounds on the variation we can achieve. In simple terms, we grow and learn, but we remain, willy-nilly, human. Even so, technologies that we adopt to extend our capabilities have changed us in ways we didn't anticipate. Simply compare the culture of today with that of a century ago. Our outlook is considerably more cosmopolitan, mostly due to transportation and communications. Our sexual mores have undergone a major transformation, due in some part to automobiles and contraceptives.

Imagine now that instead of such a minor shift in the gadgets we use, you were able to change your basic motivational structure, value system, and so forth. What if we did in fact change our perceptions so we found an office building as beautiful as a sunset? Wanting to change myself plenty of ways. If carried to an extreme, such changes, although small and clearly better at each step, could lead in the long run to people whose motives, reactions, and ways of thought we today could not recognize as human. Should we, as humans, countenance such a course, when we cannot know where it will lead?

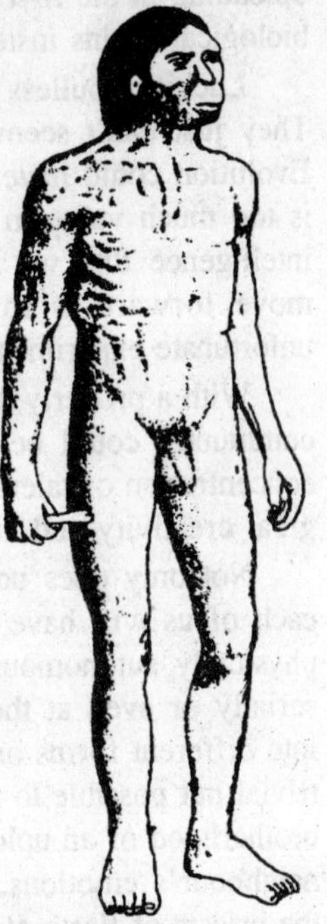

Fig. 10.6. Homo erectus.

We can no more know where nanotechnology will take us than our ancestors, *chipping stones* on the plains of Africa, knew it would lead to what we are now. Would you trade places with a *Homo erectus*? We have gained so many things: language, art, science, some would even say consciousness itself. What have we lost? Life then was nasty, brutish, and short compared to what we live today. Our present lives will seem as poor compared to what the future has to offer, if only we embrace the possibilities and work—with open eyes but also open minds—to make them reality.

It's all a matter of perspective. Either we have overrun our natural niche, jammed the globe, and have nothing left to hope for but to fight over the dwindling resources. Or we stand at the threshold of the universe, at the dawning of the age of true intelligence, and the human adventure is just beginning. The choice is yours.

11

ECONOMICS

Fifty years ago you could get a nice car for two thousand dollars. Today a nice car will cost something on the order of ten times as much. Of course, you get more for your money. Today's cars have better gas mileage, emit less pollutants, are quieter, handle better at high speeds, are safer in crashes, and last longer with less maintenance. You are more likely to have automatic transmission, and you may get on-demand four-wheel drive. That is not to mention air-conditioning, four-speaker CD players, power windows, and keyless entry systems.

The price difference is deceiving. In fact, inflation has reduced the value of money by more than a factor of ten since 1950. In other words, today's car costs about the same, in constant dollars, as the 1950s one. You are getting quite a bit more for your money.

If you look at computers, though, they leave cars in the dust. By no it is an old joke what cars would be like and how much they would cost if they had kept pace with computers over the same period. Just to get some feeling of how computer technology has changed, though, remember what a typical university computer was like when I was in college in the 1970s. We had the university records in a punched-card file that formed a twenty-foot wall in the computer center. At five feet high and a yard deep, it weighed about ten tons. There were about two hundred drawers holding three thousand cards each, and each card stored one eighty-character line of text. That less than fifty megabytes: the capacity of a stamp-sized flash chip for a camera or a USB stick today is typically five times as much.

The computer itself cost $100,000 and was bought with a grant from the *National Science Foundation*. It had 65,536 bytes of memory.

Dollars per Megaflop

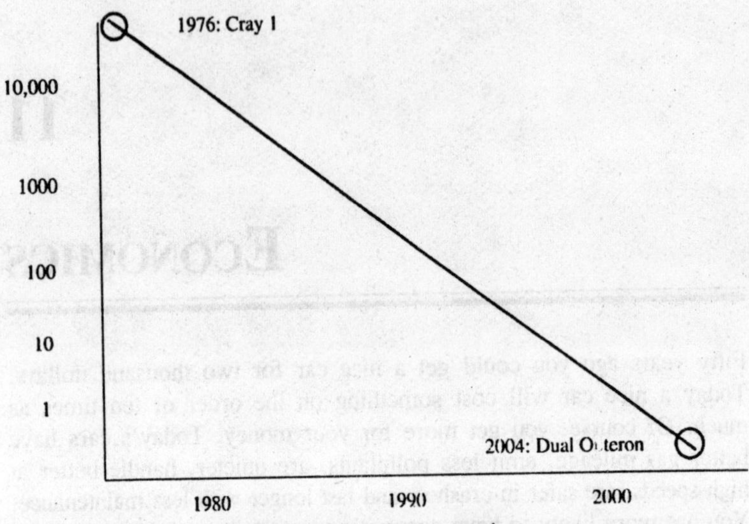

Fig. 11.1. The cost of computing.

Each bit was an actual *ferrite core*, threaded by hand. It could process approximately 250,000 instructions per second. In other words, it would have taken ten thousand such computers, at a cost of a billion dollars, to do the work of a good PC of today. And that is only in thirty of those fifty years.

Now we come to the really interesting question, which is in some sense at the very heart of the *nanotechnology*. Is it possible that over the next fifty years, the products of physical technology, motors, cars, factories, airplanes, what have you, could undergo the same kind of optimization and price reduction as computers did over the past fifty? If so, it is quite reasonable to think of owning high-tech stuff whose equivalent today would cost a billion dollars, like a large factory, an ocean liner, a fleet of aircraft, or a spaceship. You could reasonably expect to be able to carry something in your pocket whose equivalent today would weigh ten tons, like a comfortable dwelling.

The key to the price of something is the costs of its factors of production. That includes the raw materials, capital equipment, land, labour, transportation, middlemen, taxes, and so forth.

It should be pretty clear that if you have a *synthesizer sitting* on your counter, then land, labour, transportation, and middlemen disappear. Even if there were a tax on synthesized things, taxes tend to be levied as percentages of costs—they would remain in proportion. This leaves the raw material, which could be coal, natural gas, oil, or even wood. Delivered to your home in ton quantities, these materials cost from 5 to 35 cents per kilogram. If you convert them directly to a one- ton car, the raw materials cost of the car would be, say, $200. (In fact, a nanotech car could weigh much less than a ton.)

Suppose a synthesizer weighs in at 10 kilograms, a dollar's worth of coal. Suppose it takes a synthesizer an hour to make another synthesizer. In other words, if I have a synthesizer, I can make 8,760 new synthesizers in a year. Since I can borrow n at 5%, if I can sell them at $2 apiece, clearing $1 over the raw materials, I can break even if the first one costs less than $175,000.

But it would be really dumb to sell the synthesizers off as I made them. After the first hour, I have two synthesizers. If I use both of them to build more, I have four after two hours, eight after three hours, and so forth. In fact, I have over sixteen million synthesizers after twenty-four hours. At 5% annual interest, the interest for one day is 0.0137%. If I can clear $1 on each new synthesizer, I break even if I borrowed $116 billion for the first synthesizer.

Of course, no one is going to lend you $100 billion to buy a synthesizer and accept the synthesizer in repayment if you are now selling them for $2. The example just shows how an autogenous technology has the mathematical capability of driving a billion-fold price reduction.

In the real world, it might work something like this. A company invests $10 billion in making the first synthesizer. They sell ten to other companies, for a billion each, recouping their initial investment. Each of these companies sells ten at $100 million each, and so forth, $10 billion total being exchanged at each stage for ten times as many machines, until a billion machines are sold for $10 apiece, no one having lost any money.

Alternatively, the inventing company could hold the *autogenous technology* closely; selling only products of the machines and *consumer synthesizers* that could make products but not more synthesizers. Home synthesizers could be built to require a payment to the company for each object produced. Handled right, they could amass a tremendous monopoly and extract enormous amounts of cash from the public.

Since synthesizers, or indeed any form of autogenous technology; could be such a huge cash cow, any company that had it would go to extreme lengths to prevent others from getting it. This is what economists call monopoly rent; you are willing to spend a large part of the difference between your monopoly price and a fair market price to protect your monopoly, even though the resources so used do not good to society at large (and indeed generally do harm). The first place a would-be technology monopolist looks is for patents.

A patent is simply and purely a grant of monopoly. Why would a supposedly enlightened government, which has laws against monopolies in other forms, grant them? The original idea was the opposite: you wanted the inventor to publish a description of the invention instead of keeping it secret. To induce him to, you offered, legally, some of the protection that he would have gotten by keeping the secret, enough to get a good head start on the competition.

It is not a bad idea, if it were done right. But it is not. Because the patent office makes money on each patent issued, it has an incentive to patent any silly thing. Suppose the invention is obvious enough that it is easier to reinvent it from scratch than to search the patent records, interpret, and adapt the record of the patented version. Then the economic effect of the monopoly grant is a pure, unalloyed loss to society. Even if it were relatively costly to reinvent the gadget or whatever, that cost would have to be balanced against the social cost of the monopoly rent of the patent.

A side effect of patents is that when someone patents a gadget that turns out to be profitable, everyone else tries to invent a new one that does the same thing a different way, to get around the patent. So it is virtually certain that people will be trying to build autogenous technology using every conceivable scheme, architecture, and design. Not only will this lead to the deployment of less efficient schemes, but it will probably push the late arrivals toward designs that could be more dangerous, such as loose replicators.

One other thing about patents: they do require you to publish the design of the thing you are patenting. And they do not prevent people from building the thing, only from producing and selling it. Building the invention for research purposes is explicitly allowed. Patents were, after all, intended to advance science and the useful arts.

So we had expect there to be lots of *research replicators* built. Depending on the investment involved, this could be by companies or hobbyists. It is not clear whether this is good or bad—it's just something we should expect given our current legal situation.

It would be nice to think that if autogenous technology were developed by one company that produced a standard line and sold them to everyone, better control of safety and quality would be relatively easy to maintain. Unfortunately, history says differently. For example, in the 1970s, American automakers let quality slide so much that Japanese carmakers were able to make large inroads into the market using quality as a selling point (by 1980, that is; earlier the opposite was true). Monopoly engenders poor quality as well as high prices.

Another monopoly rent effect is that the would—be monopolist designs its products to lock in its customers. In other words, the products are compatible only with other products of the monopolist. A classic example is in the 1960s and 1970s, when IBM was the eight-hundred-pound gorilla of the computer world, and IBM computers used the EBCDIC character code while the standard for the rest of the world was ASCII. (ASCII remains the standard today.) IBM files looked like gibberish on non—IBM computers and vice versa. (With the loss of their dominant position, IBM systems arc now much better team players.)

At the same time, cooperative efforts of hobbyists and amateurs have built operating system software of markedly better quality, such as Linux, which is given away for free. The economics of replication are such that synthesizers, and designs for their products, could be produced the same way. Churches and other eleemosynary institutions do not give away free software today, but synthesizers are a more direct and obvious application of the charitable instinct. Synthesize someone a meal and you feed him for a day; give him a synthesizer and you feed him for a lifetime. The necessities of life, indeed for quite a comfortable lifestyle, could be had by the general population for not much more than the price of the raw materials.

Robinson Crusoe

Imagine you are on a deserted island. There is enough plant and animal life for you to support yourself working several hours a day, but you are not living in luxury by any means. Now a crate washes up on the beach and out climbs a robot. It is designed to serve humans. You are the only one around, so it's all yours. It turns out to be stronger and faster than you are, skillful, smart, eager to help, and tireless.

First thing you know, you are in a paradise. The robot does all the work and even provides some luxuries. You can do whatever you want: relax, play music, and attempt projects you did not have time and energy for before.

Now suppose instead ten people live on the island and ten robots show up. First thing you know, the people are bickering over who gets what. The robots are all working for some of the people doing what the others used to do. Some people are living very luxuriously and others are less well off than before the robots came. Much of the robots' efforts are bent toward keeping the less-well-offs from getting what the more-well-offs have.

Economics is the science of why the second scenario is so different from the first. *Money* is a remarkable invention. In a premonetary tribal setting, people traded goods but gained flexibility remembering who their friends were, who was generous and thus deserving of generosity, and so forth. Cheaters were known by reputation and soon learned that they would have a much harder time of it if they did not make amends. The problem with such a system is that it does not work for communities of more than about two hundred people. The human brain cannot learn and hold information at the rates and volumes necessary for an "*everybody knows everybody*" society much bigger than that.

Money allows for fairness in economic dealings of more or less unlimited scope. The amount of money you have, in theory, represents the amount of goods and services you (or your *ancestors*) have provided to others, minus that which has been provided to you. So if you have the money to buy something, you deserve to have it. In practice, it's a lot more complicated. Still, it works well enough to allow worldwide trade and to organize vast patterns of activity involving people who have never seen each other.

The alternative to money is the "*command economy*." In this model, some entity, be it a tribal chieftain, factory boss, people's revolutionary council, or *giant superintelligent computer*, simply tells everyone else what to do. This model works better than money for groups up to two hundred or so, assuming the boss is competent. At larger scales, it becomes less efficient. At a national scale, it is a disaster. When the Soviet Union broke up, various venture capitalists studied its industry with an eye to setting up new companies. Not much happened, because it was found that on the average, the products that Soviet industry produced were worth less on world markets than the raw materials used to make them.

Still, there is the problem of automation, made worse by the expanded capabilities of nanotechnology. Are we all doomed to be thrown out of our jobs and replaced by machines? Is some form of socialism necessary to avoid it?

In common automation scenarios, the company automates some process previously done by humans, throwing them out of work. People have worried about this sort of thing over most of the twentieth century. Some 90% of farmers have lost their jobs since 1900, replaced by automation. Manufacturing has shrunk, not as drastically but farming and manufacturing together today are only about 30% of the economy. The rest is information handling and services. There were no computer system administrators in 1900 and many fewer restaurateurs than now. Does that mean that the trend will continue, until all goods are made by machines?

Note, too, that information and service jobs are not immune to automation. *Robotics* and AI (*artificial intelligence*) are progressing rapidly, and it is easy to project a significant proportion of automation in these fields over the next few decades, *nanotechnology* or not. *Eminent roboticists* such as *Hans Moravec* predict an inexorable tide of robotic capabilities washing over the successive foothills of human ability and replacing us in the economy.

Just as a thought experiment, let's assume that nothing a human can do can't be done better by the machines. The human race as a whole should he in great shape. Collectively, we are in the same position as Robinson Crusoe and his supercapable robot Friday.

Even individually, we'd love to have our jobs be taken over by machines. Simple scenario: Someone gives each person a robot that can do his job. The robot does the work, and the person still draws the pay. No problem! The problem arises not in the existence of the robot, but in the question of who owns it. Giving each person a synthesizer has the same effect. We have seen that it's feasible to do this with nanotechnology. People could be more truly independent than we have been since paleolithic times, when we were hunters and gatherers, able to live off the land.

Drawbacks

Would we be demoralized at having our effective roles being taken over? Would we feel worthless, dependent? Humans take pride, indeed base a large portion of our personal identities and self-esteem, and tend to categorize other people, in terms of what they do. He is a carpenter, she is a reporter. In a tribal setting, such stereotypes wouldn't mean as much since you'd know more about the individual. But also in a tribal setting, that pride is an integral part of making the society work without money. The respect or disrespect of one's peers is at least as strong an incentive to work as is the prospect of material gain. And in the long run, the respect of others, not any particular amount of goods and services consumed, is the key to a life well spent and satisfying in retrospect.

Crusoe was better off with his robot because the robot did things that *Crusoe* wanted done. Psychologists, notably Abraham Maslow, have posited a hierarchy of needs that ranges from basic physical necessities like air and food through various psychological and social desirables. A salient point of the model is that the things you are interested in and worried about are at the level in the hierarchy where everything below them is satisfied, and taken for granted. But where ever in the hierarchy you are, there is always something you want. So no matter how much Crusoe's robot does, there is always something else for it to do. He could have a troop of robots, an army, a nation of robots, and he could still see things just beyond their grasp.

So with the human race as a whole. Once we have the robots, our job becomes deciding what it is we want to have done. After that, economics becomes the study of how we translate the individual desires of persons into the collective desire of the whole human race.

There is one clear, overarching danger to the whole process. It shows up in a wide variety of forms, but it can be summed up simply. We must make sure that we are the masters of the robots, rather than their slaves. The main horror of the socialist vision is that it makes people part of a machine, rather than making machines extensions of people. The same can be said for corporations.

The devil, of course, is in the details. How do we get there, to the good vision, from here, while avoiding falling into the had vision? To begin with, let's try to examine the good vision in more detail.

Utopia, LTD

We start by giving everyone a house, a *synthesizer*, and an Internet connection. We'll assume that there are corps of robots that build and maintain the infrastructure. Each person or family, is thus provided with basic physical amenities and is not obliged to work to stay alive. As an *autogenous technology*, *synthesizers* and *robots* will be dirt cheap in a Stage V nanotechnology; but land will get more and more expensive. Ultimately we are going to pack people like sardines in megahives or move into space, or probably, both.

People will need to produce something to trade in order to get land or other limited resources. They'll probably need to produce something and give it away in order to get other people's respect and attention. A current-day example is *Weblogs*, where people do the equivalent of writing a column for free, thereby getting their opinions propagated, their names mentioned, and so forth.

Virtually everything people produce will be information. The instructions for your robots are information. Entertainment produced and consumed is information. Specifications for anything to be built n your synthesizer is information.

Current-day costs of reproducing information are minimal. If you walk into a store and buy a software application in a box, it will typically cost you tens or hundreds of dollars. The actual cost of hosting a 100-megabyte file is two cents per download. With a nanoengineered, robotically maintained infrastructure this should shrink to virtually nothing. The true cost of information will be the cost to produce it divided by the number of people who want it.

Turning a ton of coal into a ton of robots is essentially a process of *imposing information*, structure, onto it. The bulk of the cost of a ton of robots should be the cost of the raw material. The information may well have originally cost a lot to produce, but it is even cheaper to duplicate than synthesizers. The only things standing in the way are monopoly rents and transactional inefficiencies.

Limited resources, such as land and energy on Earth and matter in space, can be allocated in a number of ways. Most simply in concept, each person gets a fixed amount. This is problematic in practice since people will trade, and there will be new people. You are constantly having to take stuff away from some people and give it to others, which cannot be done in a way that people will agree is fair. The far side of the spectrum is property; where what you have depends on what you had, plus what you did with it. New people get only what their parents decide to give them.

A hybrid scheme seems possible. Historically, the real interest rate, which reflects the productivity of capital, has been 3% per year or thereabouts. In a technology as productive as nanotechnology, the real interest rate might climb substantially because of the shortening of the time that it takes a unit of capital to produce another unit of capital. Think of it this way: why should I lend you money at 3% when I could buy a synthesizer and multiply my capital a millionfold instead? To compete with that possibility, you have to offer a much higher interest rate.

Another way to look at it is that the total value of all the stuff in the economy will climb faster. That means that the government, or whoever is managing the money supply, can simply inflate to cover the difference. Prices will remain stable, since there will be more goods for the new dollars to chase. And you can give the new money out evenly as a dividend to the citizens, while allowing them to keep whatever they have, can create, or can trade for. Taxes should be unnecessary.

Getting Three

Once upon a time, an *apocryphal queen* had two prisoners in her dungeon. Wanting justice to be done, but not knowing whether either was actually guilty of the crime accused, she ordered each to tell whether the other was guilty. If both maintained the other's innocence, each would get a two-year sentence. (This is referred to as cooperating.) If both maintained the other's guilt, each would serve five years. (This is called *cheating*.) If one prisoner called the other guilty while being called *innocent*, the cheater would get one year and the cooperator ten years.

From the point of view of either prisoner alone, the obvious thing to do is to cheat. He'll serve only half as much time as if he cooperated: five instead of ten if the other also cheated, one instead of two if he cooperated. But for both the prisoners combined, they'd be better off cooperating than any other combination: four total years instead of ten or more.

This little conundrum is called the *Prisoner's Dilemma*, and it is a sketch in miniature of a lot of economic interactions. The essence of it is a situation where the best thing to do from an individual's point of view is different from the best thing for everyone to do from the point of view of the group as a whole.

An important point to remember about the Prisoner's Dilemma is that it really is a dilemma. If it were always better for people to take the community point of view choice, we could have evolved to do it with complete selflessness, like ants. Instead we evolved to be flexible, and to the able to think about dilemmas like this. Nanotechnology won't solve it, nor will any other physical technology. Even if we were to genetically reengineer people to be selfless, we'd all be worse off People's individual intelligence is much better at understanding what would help them, or their small group, than the nation or world at large. A large chunk of the good of the whole really is the sum of the good of the parts. The trick, and what evolution has tried to do with humans, is to strike a balance.

The major roadblocks between us and the good vision are Prisoner's Dilemmas. Governments (or corporations or individuals for that matter) don't get more efficient without a lot of pressure. Most people and corporations involved in the development of nanotechnology will be out to gain the greatest advantage for themselves, and in many cases that will mean restricting the benefits for as long as possible to get as high a price as possible, as in the discussion of synthesizers above. Indeed, corporations are required by law to act this way—they can be used by their stockholders if they don't.

In the extreme, people (corporations, government agencies) who occupy an advantageous position in today's prenanotech world will find it in their interest to oppose any nanotech development at all. They will at least try to get the particular benefits that encroach on their current turf banned or regulated.

This monopoly rent seeking is a particularly vicious case of the Prisoner's Dilemma. If everyone in society has a monopoly, none is gaining an advantage, and all are being made poorer. If some wish to be nice, and cooperate to mutual advantage, they can do so and exclude the cheaters—in a free society. Cooperators thus have an advantage, and can prosper. Cheaters are forced to deal with other cheaters and are at a disadvantage in the long term. Thus cooperation can evolve. But in a society where cheating, that is, monopoly, is enforced by law, the only way cooperators can prosper is by operating outside the law.

Again, this is not a *problem nanotechnology* can or will solve. It is a problem that will impede nanotechnology from solving the things it could, and is another of those features of the existing world, unlikely to change, that will make nanotechnology more dangerous than it need be.

Another roadblock looms between us and nanotechnology. Many affluent Americans and Europeans are quite comfortable today; thank you. It is not politic to come right out and say. An industry has already sprung up of apologists who will try to demonize any technology that seems to have the capability of drastically improvIng the human condition for everyone.

This being the case, the *prognosticator* is also on the horns of a dilemma. The majority of nanotechnology research today is being clone in the United States and Europe. These are just the places where the currently affluent and powerful have the least need of it and where alarmist arguments are most likely to get a hearing. In plenty of places in the world, however, people are willing to take risks to improve their situation. Although they are well behind on the research curve, there is a very good chance that early adoption of nanotechnology, and thus the experience which could engender faster development, could well happen in Asia, Africa, or South America. This has a clear historical precedent in the way England outstripped France during the industrial revolution.

On the other side of the argument, we must note that rich West erners still age and die, and they are likely to support any developments that extend vigor and vitality. *Nanotechnology* could virtually eliminate physical handicaps, and ameliorate other kinds, another laudable and popular goal. We do seem determined to overeat, although we are not so fond of diabetes and cardiovascular disease. Once nanomedicine begins to make inroads on any of these fronts, it is likely to pick up steam. However, *nanomedicine* is well behind genetic manipulation at this point, and probably will he for some decades—biotechnology simply has a big head start. So we may wind up with biotech, the more dangerous of the two technologies, and not nanotech, which could have helped with detection, shielding, and alleviation of biotech threats.

12

SPACE

It was 1969, the year of *Apollo* 11 and Woodstock, the tumultuous 1960s had just seen a countercultural revolution and a widespread disaffection with science and technology, especially on university campuses where government-sponsored research was associated with the Vietnam War. At Princeton, physics professor Gerard O'Neill, inventor of the particle storage ring, had drawn the rotating duty of the freshman physics course and was looking for some way to make it socially relevant.

That was also the year that the San Francisco city council adopted the Earth Day holiday, which was then celebrated on the vernal equinox in 1970. In other words, *environmentalism* was becoming relevant. So O'Neill put all this together and challenged his students: "Is a planetary surface the right place for an expanding technological civilization?" And if not, the unstated subtext implied, what was?

O'Neill and a small cadre of interested students got to work, and over the course of the next few years, an interesting and somewhat unexpected answer began to emerge. In science fiction, the consensus view that had developed into an *orthodoxy* over the first part of the century had assumed that we would eventually settle onto the surfaces of other planets. But O'Neill's group started out, perhaps by the accident of the wording of the challenge, with a different perspective. They found an alternative: build living places in space, from scratch. And the more they looked at it, the more the numbers seemed to work out.

As the 1960s turned into the mid 1970s, the idea had caught on and the designs took on significant detail. The Club of Rome's *Limits to Growth* had come out; the energy crisis was in full cry; environ mental concern was increasing; the Earth's population seemed ready to burst the seams of this small, overburdened planet.

"*O'Neill colonies*," as they came to be called, would be miniature, inside-out worlds. They would rotate to provide gravity; and since the ground would be on the outside, it would serve as a radiation shield in the place of Earth's miles-thick atmosphere. Plans and artists' conceptions called for *lush garden communities*; lots of plants would help recycle the air. Farming and industry would flourish nearby in separate structures.

In space you can set out *mirrors* thinner than household aluminum foil, and they'll just stay where you put them, with no wind or weather and the completely predictable microgravity of orbit. Such mirrors collect a completely clean, uninterrupted, totally reliable solar power. For example, a burner on your stove in the space habitat could be powered by mirroring in sunlight from just a couple of square yards outside. No coal mining, no atmospheric effluents, no nuclear power plants, no cross-country power grid; just a collector the size of a golf umbrella.

The material to build the habitats would come mostly from the Moon, whence it was an order of magnitude easier to get into the orbits of interest than material from the Earth. *Lunar soil* was being analyzed by the follow-up Apollo missions and found to contain almost all the necessary elements—only hydrogen would have to come from Earth. High-energy smelting and other industrial processes, too costly to use on Earth, would be feasible with the bountiful, essentially free, solar energy.

As the ideas gained popularity, an increasing number not only of enthusiasts but of serious engineers at MIT, Stanford, NASA, and various aerospace companies began doing analyses and contributing ideas. (Among them, by the way, was a young K. Eric Drexler, who wound up doing his master's thesis at MIT on vacuum vapour-deposition space manufacturing.) Not only were space colonies feasible in the known technology of the 1970s, but they could pay their own way: solar power could be harvested and sent to Earth, replacing fossil fuels and obviating a host of environmental problems.

Most amazing of all, space colonies represented a flat-out, categorical solution to the problem of overpopulation. The trick is that space colonies, complete with people, can be self reproducing. That is, the people of one colony build another one, which is populated by people from Earth. Each of these builds another, and so forth. The numbers showed that by building in the asteroid belt, there was enough easily accessible material of the right and to build colonies with entry thousand times the habitable area of the Earth. (After which you can begin looking at moons, planets, comets, etc.)

In the thirty years since these ideas were developed, the Earth's population has risen from 4 billion to 6 billion. That is an annual increase of 1.4%. Increases in poorer nations tend to be higher, in richer ones lower. The United States has averaged about 1% recently. If this rate could be made to apply worldwide, by raising the standard of living, for example, keeping the population static would require moving 60 million people off the planet annually. For comparison, the three New York area airports handled 92.6 million paying passengers in 2003. In other words, three large spaceports could handle the Earth's population growth at US growth rates; at third world growth rates, you'd need six spaceports.

The bottom line is that space colonies represented solutions to quite a few of the world's problems as seen from the 1970s: population (and hunger, since the colonies are self sufficient in food); the energy crisis; pollution; and to some extent war, if fought for lebensraum. At its height in about 1977, it could have been called a movement, backed by the ten-thousand-member L5 Society and numerous public figures. We more, according to a wide sampling of leading scientists and engineers, it was within the technological capabilities of the day.

Space colonization foundered on the rocks of several realities. Largest and most jagged was the issue of the cost getting to orbit. In the 1970s, the space shuttle was on the drawing boards, and NASA confidently predicted that it would lower the cost of access to space. The cost in the 1960s, using the Saturn V moon rocket booster, which was thrown away as part of the launch process, worked out to $3,80() per pound of payload. Surely, with a reusable space truck we could do better. Then came the inevitable bureaucratic reality By NASA's own extremely optimistic figures in the 1980s (before any explosions), the cost of shuttling to orbit was $6,000 per pound. More realistic accountings by independent analysts put the cost as $20,000 to $3 5,000 per pound. Shuttle proponents had promised O'Neill that it would reduce costs to $430 per pound. Woops.

Next was the fact that a lot of the eco-angst of the 1970s turned out to be way overblown. The energy crisis was, in the medium term anyway, a product of OPEC price manipulations and not a sign of impending doom. Higher oil prices encouraged the development of new sources of supply, as higher prices will. The cartel collapsed in the face of competition, as cartels will. Although nuclear power became a political whipping boy in the United States, it was developed into a major energy source in places such as France. The "*green revolution*" and continuing improvements in firming methods and crop yields kept pace with the population and prevented any famine not caused by intentional political intervention. The predictions of *Limits to Growth* and its ilk simply didn't come true: we were supposed to have run out of oil and "many key minerals" as long ago as 1985.

And finally, even as envisioned by the would be space colonists, the project would have required a national effort of *Apollo Project* magnitude for twenty years. The political will simply was not there.

Are we, then, stuck on Earth? (Not being able to move away counts as stuck.) The answer is no, we are not stuck. What was technologically feasible thirty years ago is feasible now, and will be even more so in the future. Although we are not about to starve in the dark here on Earth, land prices do keep rising. And people do seem to keep worrying about fossil fuels. So at some point it seems quite reasonable to predict that the trends will meet to make space-based land a better buy than Earth-based.

Another point is made by a quip that was heard in the early days of aviation: "Airplanes breed like rabbits, and dirigibles breed like elephants." The bigger a project is, the harder it is to get off the ground, even with the same rate of return. So the point where space colonization really becomes a force is most likely to be when a moderate-sized group, say 250 people, can pool their resources and build themselves a new home, not unlike the groups involved in crossing the Atlantic in tiny sailing ships from Europe or crossing the West in Conestoga wagons from the East.

What kind of a dent could nonotechnology make in the problem, and how soon? Let's imagine it is about fifty years from now, that is, the amount of time that passed he Kitty Hawk and the 707. There are plenty of things, nanotechnology has not done yet, but substantial progress has been made on several fronts. In particular let's assume most of the progress is not specifically aimed at space colonization at all, but at meeting very ordinary terrestrial needs.

Foremost is *nanomedicine*. Several kinds of diseases involve intracellular damage at the molecular level, from things like free radicals and radiation. *Osteoporosis* involves calcium loss from bones, and it is likely treatable by some relatively minor chemical and electrical props at the right places in the body. People will very likely want to maintain fit and healthy bodies without having to exercise every day. *Nanomedicine* will very likely have addressed these issues.

This will make the design of a space habitat considerably easier (not to mention cheaper). The single most massive part of an *O'Neill-style colony* is the radiation shield, six feet of lunar soil surrounding the living space. With a higher radiation tolerance given by cell repair machines, a much less expensive shield provided by a magnetic field and a small shelter for major solar flares should be more than sufficient.

The other major design constraint was due to having to rotate the structure to maintain artificial gravity. *Nanomedicine* addresses this as well. It seems quite likely that the human body could be maintained in fit shape in *microgravity* with the same kinds of interventions that will already be popular to keep fit without exercising.

Avoiding both these constraints gets us out of the economic land of elephants and back among the rabbits. Much smaller habitats become feasible. Elaborate systems of mirrors to channel sunlight past radiation shields disappear. Heavy construction necessary because of the artificial gravity disappears. The necessity of building large circular structures that are rotationally symmetric disappears.

Farming and food production in space were among the easier of the problems that the O'Neill studies looked at, partly for the same reasons: plants are typically radiation resistant and not bothered by weightlessness. Even without nanotech help, space agriculture seems feasible and economical. With nanotech help, such as direct synthesis of some nutrients and some foodstuffs, assisted breakdown of waste products, and so forth, it is a very moderate item in the budget and on the chore schedule of the colonists.

In fact, nanotechnology improves the economics of the space habitat so much that it's almost viable to bring the building material up from Earth. But mass from the Moon will still probably be only 1% to 5% the cost per pound. So you'd most likely bring up the *nanomachines* (and hydrogen) from Earth, and buy raw material from the Moon.

Compared to the ten-thousand-person designs of the 1970s, the technology of 2050 will make possible single-family dwellings in space, completely self-sufficient, and easily within the economic reach of an average American family of the period, if current economic trends hold true. This might well be less than the price of a comparable house on the ground in many parts of the United States or Europe by that time. Of course by that time you'll have other options as well, from mountains to high latitudes to seaborne to submarine living areas. But even those will start to look crowded in time.

Besides the things that nanotechnology allows us to leave out of such a house, what does it allow us to put in? Windows, for one thing. Large panes of transparent material with a strength and toughness suitable for structural applications. A decided advantage will be the ability to join such panes (and other sections) with no compromise in structural integrity. Another will be that all the hull material, not just windows, will be smart. It will need to be able to change its optical properties, not so much for lighting control as to regulate temperature. It will need to check its own condition constantly, be able to do at least minor self-repair, and raise an alert when a major repair is needed.

Inside the house, you'll find the same kinds of things that nanotechnology will be providing for people on Earth—virtual reality walls or booths, robots, and the whole infrastructure of the information economy. You'll also find a much more thoroughgoing manufacturing capability than would be usual in a terrestrial home, just as you find more mechanical do-it-yourself capabilities on a farm than in a city apartment now. Apartment dwellers of today rarely need to weld a broken tie bar on a tractor; those of tomorrow will be as unlikely to need to recycle a spacecraft.

The main improvement over current practice in space, though, is the spacesuit. This is something Drexler worked out in the 1980s and described in *Engines of Creation*. It starts out a lot like the skinsuit, but thicker. It should have the same "*feels like nothing*" interface to your skin, but instead of protecting you from a temperature difference, it protects you from a pressure difference. To this end it is thicker, maybe a millimeter or so on the fingers to an inch or o on the back, forming a storage area.

The reason the suit must be thicker is that the atmosphere pressure it contains exerts quite a lot of force. Over the approximately of square meters of skin area on a human, the 100 kiloPascal pressure of air on Earth exerts a force of 200 kiloNewtons, that is, 20 tons. You don't normally feel this because it is distributed equally over your body in all directions, and your body is mostly water, incompressible for all practical purposes. But the spacesuit has to hold that much force without anything pressing back from the other side, so it will try to blow up like a balloon.

Current-day spacesuits are we difficult to move in. Take an inner tube and innate it to 15 psi, as if you were going to use it for a pool toy. Now try folding it in half. Imagine doing that everytime you moved your army you fold the inner tube, you compress the air inside to a smaller volume; release it, and it pops back to its original shape and volume. To try to help, current-day spacesuits are used with as little pressure inside as the human body can tolerate, something like 20% of sea-level pressure (that low, it has to be pure oxygen). Even so, moving and working for an extended period is quite clumsy and very tiring.

Your body moves effortlessly in sea-level pressure here on Earth because when you move, your body changes shape but not size. If you expanded or contracted, you'd have to work against that twenty-ton weight. The solution for a spacesuit is the same: maintain a constant volume inside, and there is no force favouring one configuration over another. Because your body maintains a constant volume, the suit can do the same just by hugging your skin exactly. (Breathing is a separate issue, and will require a separate solution.) But to conform to the shape of your skin, the suit needs a lot of strength to keep from blowing up into a round balloon shape. What is more, it has to maintain this strength while changing shape as you move!

A spherical balloon of diamond with two square meters of surface area would need to be less than half a micron thick to hold sea-level pressure. The hardest shape to hold pressure with is flat. A flat diamond sheet holding sea-level pressure across a foot-wide gap would have to be almost a millimeter thick—two thousand times as thick as the balloon! A one- or two-centimeter thickness over wide flat areas of the body, like the front and back of the torso, gives us plenty of margin for safety, as well as room to put lots of machinery inside along with the lead-bearing structure.

In fact, all the load-bearing structure is active. The suit is, in effect, a hollow robot that you just fit exactly inside of. It gets its commands from pressure sensors touching every square millimeter of your body, and responds so fast and precisely that you feel no resistance to your motions at all.

Besides pressure, a competent spacesuit must provide other aspects of the *Earth's environment*. The skin needs oxygen and it needs for sweat to evaporate. It needs protection from the strong ultraviolet (UV) light of the Sun, but it needs UV in moderation for vitamin D synthesis and as an input to the circadian rythm. The suit, if properly designed and programmed, can provide all of this. It can bathe you and sunbathe you, remove mites and bacteria physically instead of having to use chemicals, and even give you a massage.

For a "*helmet*," the simplest of many options is to have a flap of material in a bubble shape that attaches around the neck. It could inflate to a transparent sphere in use, and roll itself down into a collar when you go inside, like the hood of a windbreaker.

Like the walls of the house, the suit regulates internal temperature by controlling incoming and outgoing radiation (sunlight). And one of its major functions is self-monitoring and self-repair. It should be about as comfortable to wear indoors as current-clay inert clothing, so it should make living and working in a space environment as convenient as in a temperate environment on Earth.

Getting There

The major hurdle to getting into space is economic. By its, the government would have had to do the equivalent of giving each of the L5 Society's ten thousand members a million dollars a year for twenty years for the O'Neill colonies to become a reality. Current NASA launch costs are too high for economical space colonization by a factor of about one hundred. It like sitting at home on Saturday night because cars cost a million bucks. What could we do to alleviate this problem?

The first step is building space vehicles that are not thrown away each time they are used. This can be done two ways: either build a single spaceship that starts, flies, and lands as one piece, the way your car does; or build one that stages, as current rockets do, but in such a way that each piece flies safely back to be reused.

With the *space shuttle*, NASA tried in some sense to do both. The shuttle orbiter is a technological tour-de-force of the kind needed to make a *single stage to orbit* (SSTO) vehicle, burning hydrogen instead of the easier-to-handle kerosene, having rocket engines of an unprecedented efficiency. The *solid rocket boosters* (SRBs) are jettisoned in flight but parachute to the ocean and are recovered. Only the external fuel tank is thrown away ("only" though it is the biggest part). The problem is that they pushed the technological limits too much, and the shuttle needs a thorough and very expensive overhaul after every flight—and even then it occasionally blows up.

Nanotechnology will probably make a true SSTO possible. But it is not necessary to reach that technological goal to begin with. Since 1990, Orbital Sciences Corporation has been routinely putting small (one thousand Pound) payloads into orbit with their Pegasus air- launch system. This means dropping the rocket from a high-flying airplane just as Chuck Yeager's X-l was in 1947, when he first broke the sound barrier.

To get into orbit, you have to do two basically different things. You must go up to a height that will put you out of the atmosphere: 60 miles is good, 120 is better. Then you need to accelerate to about 5 miles per second going sideways. *Rockets* typically do this by starting out straight up and cu over until they are both high enough and fast enough. The engineers who design the flight paths have to juggle a lot of trade-offs. The higher the acceleration, the less time the rocket has to waste fuel just holding itself up during the ascent; but the harder it is on the occupants, and the rocket itself has to be built heavier to withstand the force. If you start fast, you add a lot of drag in the lower atmosphere, all of which has to be overcome with extra fuel.

An air launch finesses this to some extent. An airplane held up with wings can get more than twelve miles up (the U2 does, for example). At this point, you are already above more than 90% of the atmosphere and can launch sideways. You can put small wings on the spacecraft (as the Pegasus has) to let the air help hold you up while you are accelerating, so you can accelerate more slowly. That means smaller, lighter engines and a lighter frame, as well as more comfort for your passengers. Your altitude has already accounted for 10% to 20% of the height you need. Your rockets engines are more efficient because of the lower pressure. The bottom line is that only 85% of the starting weight of your rocket has to be fuel, as opposed to 95% for a ground-launched SSTO.

That makes a huge difference. If you and your family, with luggage, weigh half a ton, and your spaceship, including cabin, fuel tanks, rocket motors, wings, and all, weighs a ton empty; then loaded and fueled, you are ten tons. For comparison, the Boeing 737, a relatively small airliner, has a payload of fifteen tons. A drop aircraft is thus easily within current capabilities, much less nanotech ones.

How much is this going to cost you? With a nanotech factory, you can take hydrocarbons like natural gas and use the carbon to build things, while separating the hydrogen for use as fuel. You can also take carbon in a more concentrated form such as coal. Current commodity prices for natural gas are less than 50 cents per kilogram, and coal is about $30 per ton. Coal is not pure carbon, so we might need two tons for the one-ton spaceship. We need 50 kilograms of hydrogen, requiring 200 kilograms of natural gas. With a markup factor of one hundred for the spaceship, it costs $6,000. Liquid hydrogen today costs something like $10 per kilogram, and liquid oxygen about 10 cents. We need 50 kilograms of the former and 800 of the latter, for a cost in today's prices of $580 per flight. I'd expect this to become lower since hydrogen is likely to become a widely used fuel over the next few decades, with competition driving the cost down.

The actual price of spaceships will consist almost entirely of amortized development costs, with raw materials counting for very little. Engineering costs for new rocket engines call be in the $10 million range today. (Note that by this point the *liquid-fuel rocket engine* will be more than one hundred years old—great works of innovative genius will not be necessary.) Assume engineering for the whole vehicle at $100 million: if you can sell a million ships, you need only add $100 apiece; if you only sell 100, you'll have to charge $1 million each. The bottom line: if enough people are interested, cost is not a problem.

Longer-term Options

Once there is a reasonably high volume of traffic to space, costs can be reduced even further. One of the schemes you'll probably hear about is called the "*space elevator*," "*skyhook*" or even "*beanstalk*." You'll hear about it in conjunction with nanotechnology, even nanoscale technology as done by fullerene chemists, because it is a really nifty, compelling idea that is unfortunately not feasible with any material except perfect diamond fiber or fullerene nanotubes (bucky tubes).

Here's the nifty, compelling idea. A satellite placed in an orbit of radius 42,164 kilometers (i.e., 22,000 miles above sea level) will orbit the Earth in exactly twenty-four hours, and thus can appear to hang stationary above a given spot on the equator. This fact is much used by communications satellites today; the orbit is called geosynchronous or Clarke orbit, and it is already somewhat crowded with satellites.

Now imagine that you are standing on a geosynchronous satellite, and you let out a twenty-two-thousand-mile-long string. Someone could tie a package to it, and you could pull it up, and bingo, you have put a package into orbit without using any rockets. To fact, if you put the string over a pulley and let one package down as the other comes up, you don't need to spend any net energy at all! Skyhook concepts typically are a lot more involved, with elevators, and counter weights on the satellite to balance the weight of the string, and so forth, but that is the basic idea. And what is more, in theory; it would work.

That's in theory. In practice, the thing is huge, if narrow. If it breaks, the cable comes whipping down into the atmosphere like a twenty-thousand-mile-long meteor, causing widespread consternation. Probability of breaking: 100%. satellite that is not in *geosynchronous orbit* is guaranteed to hit it, eventually at speeds on the order of five miles per second. According to the US Space Command in 2000, 8,927 human-made objects were in orbit. Even if we cleaned these up and prohibited any new satellites, problems remain. Because the skyhook has to be near the equator, it goes through the Van Allen radiation belt. You are going up at elevator speeds, not rocket ones, which means you get to soak up quite a lot of radiation on your way up. If the elevator runs at 100 mph, it takes more than nine days to get to geosynchronous orbit.

Rockets breed like *rabbits*; skyhooks breed like *elephants*—and the two methods are mutually incompatible. By the time technology and economics make skyhooks possible, there'll be too heavy an investment in satellites to change course.

A somewhat more realistic idea can be had by looking at ships. Ships move well in water, but have a pretty hard time on land. Typically when you need to get on and off a ship, you do not beach it, but bring it up to a pier. A *pier* is an extension of land built so that a ship can operate in contact with it without having to be beached.

How do we build a pier for spaceships? The equivalent of the beach, for Earth, is the atmosphere. If we build a tower twelve miles high, it gives us the same advantages as the air launch. If it is long, as well as high, we can help the rocket by putting a mass driver on top. A mass driver is essentially a cannon that uses magnetism instead of an explosion, so the "*cannonball*" can be something relatively fragile, like human bodies.

How much can we help? In fact, if we make the tower 60 miles high, and 180 miles long, we do not need the rocket at all. You are above all but one-millionth of the atmosphere. Accelerating in the mass driver at 10 Gs, an acceleration healthy humans can handle, will put you into orbit. Sixty miles is high enough for orbital speeds in the mass driver, but low enough not to bother satellites. A 100 mph elevator ride to the top of the tower takes only thirty-six minutes. The launch through the mass driver takes less than a minute and a half.

For freight, we don't have to stop at LEO (low earth orbit). Mass drivers delivering 30 Gs were built in the 1970s. Thirty Gs on a 180-mile track gives the ship enough oomph to get to Venus or Mars; the

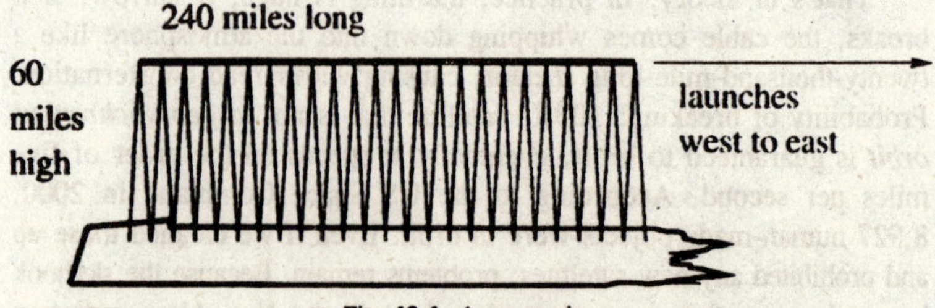

Fig. 12.1. A space pier.

Moon and L5 require only 20 Gs. People would have to make a stop in orbit to be accelerated again on a much longer driver.

Whether the tower will be worth building depends on the volume of freight to space and the cost of electricity. Running the mass driver continuously (and assuming ten-ton spaceships) would put nearly four million tons of shipping into space each year, and require 4 gigawatts continuous power. *Nanotech-based solar power satellites* can probably he designed at about 10 megawatts per ton, so the tower needs to launch forty for its own use—that will take an hour. The tower itself costing less than $10 billion, so they would represent a trivial fraction of its cost as well. If the tower did cost $10 billion, and was used at full capacity, amortized cost would be under $100 per ton, which is comparable to current international seagoing freight rates. Note that over five million tons of freight is shipped annually on the Snake River alone; most major US waterways handle one hundred times as much.

13

TRANSPORTATION

In some parts of the country, like Alaska, private airplanes are more common. Planes have a couple of disadvantages compared to cars, though. They are expensive. They are less intuitive to operate and require a lot more training to use safely. They are more susceptible to bad weather. And they require a lot of space for runways and clearance from other airplanes. Still, we fly routinely on airliners that go, on average, ten times as fast as cars. They are safer than cars, and cheaper for cross-country travel.

But traveling in airliners is a major pain. Consider a typical trip in the middle distance, say, five hundred miles. First, you do not get to pick when the flight will be; it may match your schedule, but in most cases you have to modify or plan your schedule to match the airlines. Then you drive to the airport and park, which takes an hour. Then you check in and wait, because you have been advised to be at least an hour early for the flight. You cannot even carry a pocketknife. Everything you take has to be packed for each trip and will be inspected by officious strangers. Once the plane leaves the ground, every thing is smooth until landing an hour or so later. Then it takes another hour, or more, to get your luggage, rent a car, and drive to your actual destination. And that is if everything goes smoothly, your luggage is not lost, the flight is not delayed, and so forth.

With a car, you just toss your stuff in the back and go. You can keep things in your car all the time rather than having to pack them for each trip, if you like. No poking, prodding, or searching—and no chance of being hijacked, either. No lost luggage. You can smoke if you want, play loud music if you want, or have complete quiet if you want. Stretch out, rather than being packed like a sardine. Stop when you like and eat in a nice restaurant.

Aircars

Since the 1950s, people have been tantalized with the idea of private air cars, which held out the promise of combining the best parts of both these modes of travel. If you could fly directly from your driveway, you could not only have the privacy and scheduling advantages of a car, but reduce travel time from four hours to one by avoiding all the inefficiencies of the airport. Even if your air car is only half as fast as an airliner, you'll save a couple of hours.

Could nanotechnology build an *aircar*? There is no doubt what so ever that it could, because current technology could build an air-car. The salient questions are whether it can reduce the cost, increase the safety, and make it quiet enough to use in residential areas without the neighbours getting-up in arms.

Current aircars, such as the *Moller*, are listed in the neighbour hood of a million dollars each. Mass production could bring this down considerably, but it will take something like *autogenous molecular manufacturing* to make them really affordable. Given the strength-to-weight ratios of nanotech materials and the power-to-weight ratios of molecular engines, a powerful, capacious air-car need only weigh a few hundred pounds. This means a few hundred dollars in raw materials costs. In the long run, once development costs are amortized, an aircar could be significantly less expensive than ground cars are today.

Airplane

Flying a light airplane can be a tricky business. The most common single cause of major accidents is euphemistically referred to by the FAA as "*controlled flight into terrain*," that is, diving into the ground when you thought you were flying level. It is perfectly possible to be flying upside down and not realize it, or to pour a cup of coffee while the plane is doing a complete 360-degree roll. Safe flying takes a lot of training and practice.

Flying a vtol, a craft that does vertical takeoffs and landings, is even harder. Yet we want our aircar to be a vtol. The reason is that a car-sized vtol would be compatible with existing driveways, but a "*normal*" rolling takeoff plane would require you to build a runway. (Don't even think of a separate airport away from your house: it reintroduces all the airport congestion problems and requires you to have a separate ground car, not to mention the problem of getting a ground car at your destination.)

Thus the aircar needs to have an autopilot capable of doing all light operations. These not only exist, but have been routinely in use on commercial airliners for more than a decade. An autopilot has not only the right reflexes fur the conditions of flight, counterintuitive to humans, but is directly connected! to an array of sensors that tell it a lot more about what is going on. For example, it could sense directly all the patterns of pressure and airflow around the craft. Nanosensors and nanocomputers for such an autopilot are simple for nanotechnology, whereas their current-day equivalents are expensive and high tech.

The dual constraints of vertical takeoff and automobile size make it quite difficult to design a machine that is quiet. Helicopters are obnoxiously noisy, and directed-thrust jets, like the Harrier, are ear splitting. In simple terms, to go up you have to throw a column of air down. You can throw a thin column fast, like the Harrier, or a wide column more slowly, like a helicopter. The faster, the noisier.

The air thrown down by a helicopter, by itself, is like a strong wind, and you get a rushing sound but nothing overly loud or obnoxious. What makes the noise in a helicopter is the *blades*. They produce the concentrated *whump-whump-whump* you hear because they give the air a hard kick each time they come around, rather than moving it smoothly and evenly. The noise a helicopter makes, by the way, annoys the average person even when it some 20 decibels quieter than the threshold of annoyance for smoother more continuous sounds (like traffic). It something like rock music in that respect (for some people!).

The quietest way to take off, however is to jump. It is well within the capability of *nanoengineering* to design a leg that will extend to fifty feet in length and fold up into a pad less than an inch thick. This gives us two advantages. First, the aircar is already five stories up when the air-moving thrust has to take hold, so the noise is farther away. Second, it already moving at a good clip, so less acceleration is needed. It also turns out that moving faster makes aerodynamic thrusters (propellors, ducted fans, or jets) more efficient; you are throwing down a longer column of air and thus do not have to push it as hard.

A Harrier or *helicopter* landing is almost as noisy as a takeoff. There is a way to land vertically that is virtually silent, though: with a *parachute*. Taking that as an inspiration, we can design the next stage of our vertical flight. Unfurl a sheet of thin material over the area swept out by a helicopter's rotor. But rather than cloth, this is mesh with holes about like window screen. In each hole is a tiny fan blade. Now you can move a large column of air without any whump-whumps. Instead of hanging by strings like a parachute, it will probably be more controllable to deploy the fan-cloth from spars like a sailing ship. It could make for a fairly romantic vision: sailboats in the sky.

Once you get up to a decent altitude, a few thousand feet, you need to retract all the legs and sails and move fast. The first thing you do is extend wings. It perfectly possible to stay up and move forward on thrusters such as ducted fans, but since you want to be moving fast anyway, wings make things more efficient. Typical airplane designs today have a lift-to-drag ratio of about ten. That means that the engines have to produce only one pound of thrust for each ten pounds of weight; the wings do the rest.

Virtually any current thruster-style will work, in particular ducted fans; but there may be a way to do better. An object in motion tends to remain in motion; if your aircar were in outer space, it would not need any thrust to keep going. In the atmosphere, however, there is drag. Drag is unusually broken down into three categories, which correspond to the modes in which the energy is dissipated into the air around the plane. First is induced drag, which is associated with the lift. This results in the downwash of air behind the plane. It could be lowered by having longer, thinner wings like those of a glider. They however, would increase the next kind of drag, skin drag. This is essentially friction of the plane's surface with the air moving across it, which heats the air. Third is form drag, which results in turbulence in the air. It can be reduced by using a streamlined shape.

Rigid metal parts, as in current airplanes, an approximate streamlined shapes, but the streamlines change with every change in speed, attitude, wind, turbulence, and so on. Soft, compliant wings, like an owl's, can match the streamlines much more closely, so an owl can fly with virtually no turbulence and be totally silent to human ears. *Nanoengineered wings* will be able to adapt to the airflow closely and reduce drag and noise thereby.

Skin drag happens because surfaces are not really smooth. When an air molecule hits a moving surface, the surface tends to bump it in the direction the surface is moving, transferring momentum from the craft to the air. Atomically smooth surfaces could help, but even they are bumpy and atoms are sticky. Some momentum transfer will remain.

Suppose, however, you could put a pickle wheel or propellor on the surface and give the bouncing molecule a hump in the other direction? Then you'd have negative drag. Then you could afford to

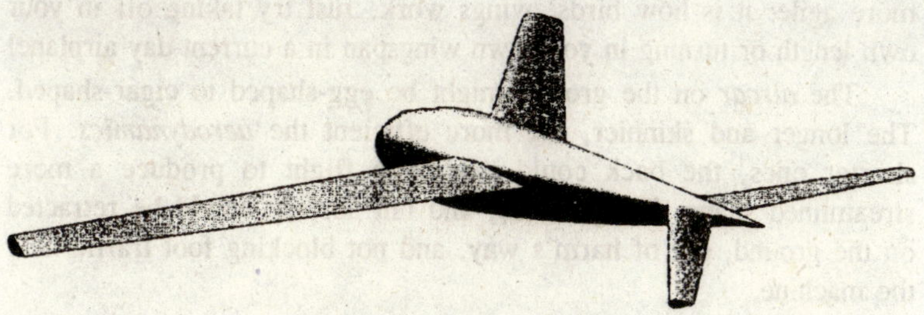

Fig. 13.1. An aircar cruising.

put out long, glider like wings, because the drag is not pulling you back, it is pushing you forward. So you can decrease the thrust necessary by a factor of five, since top-notch gliders can get lift-to-drag ratios of fifty. You have cut way down on induced drag, and get your thrust from negative skin drag.

If all this could be made to work, the power you need to fly would be 2% of your weight times your speed. Let's say 5%, to allow for inefficiencies in the negative drag, the power conversion, and so on. Even so, if you and your luggage and your plane together are half a ton, your 500-mile trip at 500 mph will cost you less than two gallons of gasoline (or less than five pounds of hydrogen).

Could it be made to work? Everything except the negative drag seems straight forward with the capabilities of nanoengineering materials and machinery. Shape-changing is too heavy and expensive for all but a few military aircraft today, but it is no problem for molecular technology; note that your body moves, and birds fly, by changing shape.

For shape changes such as extending legs, folding and unfolding wings, and extruding spars for sails, we can build structural members from very thin sheets of diamond. The layers are dadoed to each other so they would not pull apart and can only slide one way. These would be pulled across each other by millions of microscopic motors. A leg, for example, would simply be a set of telescoping cylinders like a radio antenna. It would just happen to be six thousand cylinders, each 5 microns thick, an inch long, and ranging from 6 to 8 inches in diameter. When retracted, it would form a doughnut an inch thick; when extended, a tube 50 feet long. Inflated to ordinary tire pressures, the pressure alone would lift over half a ton.

The brothers Wright steered their original Flyer by a process they called *wing-warping*. Alerons, flaps, slats, and so forth came along later. Wing-warping is a bit cleaner, less draggy, and a lot more agile: it is how birds' wings work. Just try taking off in your own length or turning in your own wingspan in a current-day airplane!

The *aircar* on the ground might be egg-shaped to cigar-shaped. The longer and skinnier, the more efficient the *aerodynamics*. For shorter ones, the back could extend in flight to produce a more streamlined shape. Wings, sails, and tail surface would be retracted on the ground, out of harm's way, and not blocking foot traffic near the machine.

If negative drag works, control surfaces could be even more minimal. Controlling the force (and possibly direction) of the drag could give you a ignore direct handle on the airflow patterns around the craft. To increase lift on a wing, make the drag on top more negative and that on bottom more positive. To turn left, use positive drag on the left wing and negative on the right. If it worked really well, you should be able to fly a simple particular shape ("*flying saucer*") whose surface had programmable drag.

The reason that negative drag is not as straightforward as, say, shape changing is that it involves using nanomachines directly in an uncontrolled external environment. Tiny just particles could smash your tiny wheels or propellors. What would be unnoticeable levels of moisture could immerse them in water. It might not be possible to find a design that overcomes all the difficulties and still gets the efficiency advantage we had like. At this point, my analysis indicates a better than even chance it could be made to work, but it is not as definite as many of the other designs. If not, we had just fall back on ducted fans, like airliners use today. Not nearly as elegant, but we know they work.

What about safety? The main causes of problems light aircraft can have are pilot error, weather, turbulence caused by large aircraft, and engine failure. *Pilot error* is minimized with the highly integrated autopilot and sensor system. Since it can communicate directly with other aircraft and radar network there is less chance of unexpected encounters. Since you are not operating near big airports, you can more easily avoid large planes. Inclement flying conditions are much more easily handled when the craft has a factor of ten or one hundred of power to spare compared with current models. Current-day air craft are tossed around like leaves by high and *turbulent winds*, but nanotechnology will make it possible to put a much higher peak power capacity in the plane without making it too heavy or cost too much. That will improve safety. And finally nanotech gives you the option of using millions of tiny motors instead of one or a few big ones. Any given engine failure would not even be noticed.

When the air traffic control system be overburdened? Yes, it would. because it is an antiquated, bureaucratic mess. If you had to ask a "*ground traffic controller*" for permission at every turn when you drove, driving would be a similar mess. Instead we have autonomous control, with automated arbitration like traffic lights, and it works reasonably well. Watch a flock of birds land in a field and take off again. They are flying within a wingspan of each other at speeds equal to city automobile traffic, with nervous systems considerably less complex than yours. They don't bump or crash. Watch a cloud of flies around a garbage can. They perform the same feat with nervous systems your PC can match. No central control in either case. Purely distributed, autonomous navigation based on local information. If, with all our efficiency technology, we cannot do as well as flies, I'll eat my hat.

What about crowding, congestion? Over cities, certainly; so what is new? Outside them, we can estimate congestion by noting that there are 200 million cars in the United States, and 4 million miles of roadway. If we spaced the cars evenly along the roads, they'd be about 35 yards apart. Since there are about 50 million cubic miles of flyable airspace over the country, the same number of air-cars evenly spaced would be over half a mile apart. That is if they were all up there at the same time.

Each American car is driven, on average, 10,000 miles a year. Leaving out cities again, let's assume this is at an average 50 mph and takes 200 hours, less than 2.5% of the time. If we assume the air-cars are traveling ten times 'as far on average, they are still taking only the same amount of time, so the average number in the air is about 4 million and they are over 4 kilometers apart. In other words, if you flew from New York to San Francisco completely obliviously, not watching for traffic at all, there would be one chance in 100 million you'd hit another car. Add radar, transponders, and an always-alert autopilot, and it is something I would not worry about.

Over densely populated areas, you (or your autopilot) would probably have to file a detailed flight plan with some central authority, essentially taking out a reservation on each successive spot in the air space you intend to occupy. This could be completely automatic, indeed completely decentralized, as traffic on the Internet is now.

Aircars, like most other powered machines operating in Earth's atmosphere, will probably use hydrogen for fuel. Hydrogen, as we have seen, is considerabiy lighter for the energy it gives you than hydrocarbon fuels like gasoline. That is useful in a flying machine.

How fast will it fly? From 1920 to 1970, the speeds (and altitudes) of commercial airliners increased along a smooth exponential curve. Then they stopped. There are two reasons for this. First is that bigger *aircraft* can fly faster and more efficiently, but with the 747, diminishing returns set in as it gets more inefficient on the other side of the economic equation to make more and more people go to the same place at the same time.

The other reason is the sound barrier. The technology to fly faster than sound is half a century old now, but flying just over the speed of sound is three times as expensive in energy as flying just under it. So modern airliners fly just under the speed of sound and right at the edge of the *stratosphere*, six miles or more up.

It is quite possible to have a supersonic private plane. Several people do today, but as you can imagine, they are hellishly expensive. But the flight regime of the current-day airliner is a high point in efficiency. We can reasonably expect the design of the everyday flying car to follow the same logic.

At airliner speeds, Punxsutawney, Pennsylvania, is as close to New York in time as central New Jersey is now. And closer to Philadelphia, Pittsburgh, Cincinnati, Cleveland, and Toronto. Perhaps Phil the groundhog will have some new neighbours. If the 1-78 phenomenon is repeated, and it is by no means an isolated case, the value of land over much of North America outside of existing metropolitan areas could rise considerably. That is the sign of a huge social benefit.

There remain reasons to want to go faster than airliner speeds. Suppose you want to visit Australia. From New York, it is twenty hours in the air and a bit more with stopovers. The Pacific leg is at the edge of the range of *modern jetliners*; once was on a light that had to put down in Fiji, unscheduled, to refuel because we had fought headwinds on the way.

The *space shuttle*, of course (or even John Glenn's *Mercury capsule*), travels that far in under all hour. The frustrating part of it is that it takes no more energy to get into orbit than it does to drag a plane through the atmosphere halfway around the world.

A passenger rocket in tile *nanotech era* might look something like a modern-day airliner. Like any new technology, it would be commercial and concentrated at first, just as railroads preceded private cars. On the inside, though, would only be a first-class cabin; the rest would be fuel and oxygen tanks. The rocket would take off on a runway (we have got all these airports already anyway) and climb just like a current-day jet of ducted fans. Instead of leveling off, it would keep climbing. At takeoff, the rocket would have only hydrogen fuel aboard. As it flew, it would suck in huge quantities of air and liquify the oxygen for use in the rockets.

Nanotech could make *rockets* quite a bit safer and cheaper than they are today, but not too much more efficient; they are reasonably near the physical limits today. There are several possible ways to get the rocket effect besides the standard combustion chamber and nozzle arrangement. One way might be to burn the fuel to produce power and use that to accelerate the reaction mass electrically. Spaceborne ion thrusters work that way today, but don't provide enough thrust in their current form. It might be possible, burning hydrogen in oxygen, to accelerate the resulting water molecules without ionizing them, since they have enough of a polar electric character to grab electrically.

Most of the *nanotech applications* would not push the technology to its limits, so it is possible to give safe predictions by using designs we can currently simulate and analyze. Rockets do push the limits, as do any of the other possible schemes for reactive thrust. So we can't say with great assurance how they will work in detail— there are a great many ingenious innovations to be made. However, we do know how to make rockets today, so we know they are possible. We might make thousands of tiny rocket engineers and be safe against any one failing. Alternatively, scaling laws favour larger ones. The best engineering compromise remains to be seen.

Your rocketliner accelerates up on fans for two or three minutes, loading its oxygen tanks. Then it cuts in the rocket engines and retracts its wings to a highly swept, sharp-edged shape; by this time it is at least ten miles up, and the noise, rockets and sonic boom, would not be nearly as noisome to people on the ground. You accelerate at less than two Gs for about ten minutes, coast fur half an hour, brake in the air another ten minutes, and fly down to the airport.

How much would your ticket cost? One way to estimate is that current-day airlines operate at about six times fuel costs. For rockets today, the technology costs much more proportionately, but nanotechnology could bring that down into line with current jet technology. Assuming your share of the weight is a quarter ton, you need the same weight in hydrogen, at a current cost of about $190 in gas form. There would he traffic at $1,200 a ticket but not high volume. If fuel costs got down to the level of coal as delivered to utilities, your ticket would cost $200 instead, and there would be plenty of takers.

Freight

The cost of coal is illustrative of a more mundane aspect of transportation. Coal at the mine mouth in Wyoming costs $5 per ton. Delivered to utility-generating plants by the train car, it is about $30 per ton. Delivered to your house in Maine by the single ton, $135.

Railroads are an old, established, highly optimized means of *transportation*. Nanotechnology could squeeze some more optimization out of the current hardware, by such means as self-maintaining track, faster trains, and smoother rides.

In the long run, though, surface transportation can only get more expensive, since it requires land. Whether train or truck, surface vehicles put the noise and danger right where people want to live. The same technologies used in the private aircar should be able to make *airtrucks* for payloads of up to ten tons or so, which would not be too much noisier than current-day trucks. Airtrucks should be at least as energy efficient as ground trucks for long distances as well. Trucks use a large proportion of their energy stirring up the air, and a somewhat smaller proportion breaking up the pavement. Carefully tuned aero dynamic shapes flying over hills, valleys, and stop signs would lower energy costs.

A large part of the need for bulk transportation could simply be obsolete with the advent of general-purpose synthesizers, or even special-purpose ones in local stores. There has been a historical trend in this direction. When I was young, I took a roll of film and mailed it off to Rochester for processing, getting the prints back a week later in the mail. In the 1970s and 1980s, the technology for one-hour photo shops in mails and main streets was available. Now, of course, you take a picture with a digital camera and print it out if you want, or simply e-mail it without its ever touching paper.

Being able to review and save pictures in digital form has cut down on the amount of photo paper necessary, but some paper still has to be manufactured and shipped in bulk. Indeed, the computer and office printer, as well as the copier, have dramatically increased the amount of paper we have to contend with. What is missing, that *nanotechnology* could provide, is local recycling. Once the waste stream can be turned around locally, by home recyclers, long-distance transport of bulk materials (not to mention waste!) will begin to decline.

Beam Me Up, Scotty

What about matter transmission? To be more specific, using *molecular dissasemblers* to take an object apart, noting the type, position, and bonding of each atom, and transmit the information to some where else, whereupon a molecular assembler puts together an exact copy?

Don't hold your breath. Let's suppose the high bandwidth link to your house is a cable with a core that consists of one million optic fibers, and that the transducers are able to send data at optical frequencies 100,000 times faster than today's gigabit fiber. Your total data rate is 100 exabits per second (i.e., 100 billion gigabits). That is a reasonable expectation for a *nanotech data link* of the later twenty-first century. It could transmit the entire contents of today's World Wide Web in a tenth of a millisecond.

The human body contains something like *seven billion billion billion* atoms, and to describe each one's type, position, and bonding you need something like 100 hits, even with compression. In other words, the time it would take to transmit a complete atomic-level description of yourself along your nanotech data link would be about 222 years. You might as well walk.

You could save a lot of data by transmitting a close, but not exact copy. Your body contains a lot of water; no need to transmit the position of each molecule, just say "*water is here*" and some notations about how salty it is, and so forth. Furthermore, many of the complex molecules are the same, or should be proteins and DNA come as many copies of many fewer molecule types, and variations are injuries you'd just as well correct. It seems likely that if you allowed a little fixing-up, and did not mind as much variation in atom positions as would happen in a tenth of a second or so, you could probably transmit the information required in a reasonable amount of time.

There is one caveat about transmitting yourself this way, though: it's just as possible to make o copies, or even a hundred, at the far end as one. Heaven only knows what cans of worms we could open! But personally, I'm not sure I'd want to transmit a fairly good copy in the first place. Let's just say that teleportation is an if proposition: not obviously impossible, but not clearly practical, either. On the other hand, transmitting the descriptions of manufactured objects will be routine. These will be hierarchically structured, built of many copies of identical parts, and much, much simpler than raw natural objects. Transmitting *full-sensory telepresence* data will be easy also; you can get close to that with a gigabit link today, and 10 gigabits is almost certainly enough for complete fidelity at human sensory resolution.

Physical personal travel for business might then decline, because high-fidelity teleconferencing could be just as acceptable as real meetings, and a lot faster. Personal tourism might well increase, though. Bulk freight might decrease but personal items—handicrafts, mementos, souvenirs, and other objects whose value lies in their identity—might increase. The mail bot will still bring a pound of unwanted paper trash each day but you will just toss it into the recycler.

14

Companion: A Personal Computer

Compact, powerful communications devices have been described in both science fiction and serious speculative technical writing. The twentieth century, with the advent of inexpensive, *miniature electronics*, has glimpsed possibilities of personal communications. The twenty-first century will dawn wit' the development of nanotechnology even more, and devices such as the communications technology will accelerate even more, and devices such as the Companion, described here, will become possible.

Purpose and Scope

The *Companion* is designed to provide information, communications, and entertainment of unprecedented quality and quantity. It will include an extensive library, comprehensive two- way communications, high-quality audio and video displays, and powerful computer facilities.

Molecular-scale manufacturing will afford a tremendous advance in data storage capacity over current integrated circuit technology. It will be possible to create a pinhead-sized library with a capacity of several million trillion data bits. The library in the Companion will contain digitized versions of millions of books, thousands of musical recordings and motion pictures, interactive course ware on many subjects, and several special-purpose knowledge bases to help us in everyday life.

Design Characteristics

1. Personal empowerment, via communication, entertainment, information.

2. Lightweight, unobtrusive, attractive, durable, convenient.
3. High memory capacity, redundancy in communications and power supplies.
4. Several easy-to-use interfaces.
5. Realistic audiovisuals.
6. Trustworthiness through privacy and security.
7. Low cost and wide availability.
8. Credible as a future product; an outgrowth of currently available technology.
9. Easy to understand, not dangerous or threatening.

Appearance

Packaged as an ordinary looking pair of eyeglasses, the companion uses the power of *nanotechnology* to deliver the services listed. The lenses of the glasses contain an imaging system to present a high-quality display. Integral earphones provide audio. Built of advanced materials, the Companion will be though, durable, and lightweight. The Companion, like present-day glasses, will be available in a wide variety of styles and colours.

Technical Details

The following sections detail the fundamental architecture of the companion.

Frames

The frames of the Companion have ample room for data processing and power supplies. Data is stored and manipulated in a few cubic millimeters of nanoelectronics and nanomachines. Power supplies will occupy up to 90 percent of the available interior volume of the frames and more than 50 percent of the available exterior surface area.

The frames are likely to be a dark collect solar power effectively. The surfaces touching the skin have a slightly sticky material that gently clings to the face. This prevents the annoying sliding- down-the-nose problem—without invoking the parts-that-pinch problem. The inside surfaces have biosensors to pick up pulse, skin temperature, and blood chemistry via perspiration.

Audiovisuals

An imaging system embedded in the eyepieces produces extremely high quality pictures. Images produced using PAO (*phased array optics*) can be indistinguishable from reality. Conventional imaging would put the picture too close to your eyes to focus properly, but PAO can create an image that seems to float a meter or two ahead of you, allowing your eyes to focus on it naturally. The imaging system is mounted on a thin film inside the lenses. It rolls up like a window-blind to permit unobstructed vision and clear eye contact when special imaging isn't needed. PAO, when in place, would reduce incoming light about as much as a pair of sunglasses.

Not everyone will need corrective lenses to see clearly, and twenty-first-century medical science may have solved vision problems permanently. Depending of the owner's needs, the eyepieces can be either optically neutral or lenses.

Mounted right above the eyepieces are two *video cameras*. They are the same distance apart as your eyes to give recorded images a greater sense of realism. The cameras can be tuned to pick up infrared and ultraviolet as well as visible light. They can also be designed to provide light amplification for use in near-darkness.

Earpieces are provided for each ear, which resemble hearing aids embedded in *eyeglass frames*. The technology for the earpieces doesn't need to be much more sophisticated than in present earphones, though they will sound better and use less power. Miniature microphones placed near the earpieces pick up stereo sound from the outside environment. As with the video cameras, the microphones can be designed to perceive a greater range of frequencies and intensities than our ears can.

Computing

The Companion employs both neural networks and conventional logic systems for computing. *Computer* neural networks duplicate some functions of natural nervous systems and are well suited to learning and recognizing patterns (a face, a sound) and linking one piece of information to another. These don't need to be as powerful of as well organized as brains to be useful in a specific role.

In the Companion, neural networks analyze torrents of broadcast and sensory data and alert other components when something interesting comes along. Serial logic systems are like today's familiar computer. They complement the neural networks by being precise and reliable. They handle internal housekeeping, library, security, and other functions. The computers in the Companion are tiny: depending on requirements, hundreds to thousands of them may be employed.

Nanotechnology will enable a gigantic library to fit into a tiny space. Data storage schemes, such as coded polymer chains, may be able to achieve storage densities of a billion bytes per cubic micron, or one million trillion bytes per cubic millimeter. This storage density is roughly comparable to DNA, which packs about 100 megabytes into a small part of a cell. The Companion's design assumes a one-cubic-millimeter block of such memory, however, if this isn't enough, the frames have room for several cubic millimeters.

This *data storage capacity* is astronomical by today's standards, exceeding the present combined storage capacity of every computer in the world. The problem with capacity advances, however, is that the marvelous gains have been temporary; people are quick to find new uses for expanded memory capacity and fill it quickly. The libraries and software described here are based on present-day trends; make these projections seem naive. The components described will occupy less than a tenth of one percent of the Companion's memory, leaving the rest free for a lifetime of use.

A self-integrity system consisting of dedicated computers and software keeps tabs on the Companion's battery conditions, structural integrity, available memory capacity, and security status. It will automatically alert a network-based emergency center if it of its owner comes to harm. If it is lost or stolen, the personal data contained inside can be automatically erased or transferred to another Companion.

Communications

The Companion uses both optical and radio channels of communication. The outside surfaces of the frames are lined with dozens of optical send-receive units; a radio antenna 20 centimeters long is imbedded within the frames. The light-based link is by far the faster medium, capable of sending and receiving several billion bits per second, but it will be limited to wherever you can find an optical outlet. Radio is more widely accessible, but slower. Because of noise, power, and distance trade-offs, radio channels will have a much smaller bandwidth for transmitting, less than 500 kilobits per second. This will be sufficient for voice transmissions or small data files.

In the next ten to twenty years, *fiber optic outlets* will likely replace conventional telephone and cable TV in most homes and offices. If you're at home or in a public building, your Companion will have the full high-speed access that optical fiber offers. Optical outlets placed on the walls and ceiling could bathe most of the room's volume in low-power infrared light. These outlets would be on all the time, When you walk into a room, your Companion would sense the infrared signal and send its own signal, establishing a two-way link. The

Companion's frames will have optical receivers all around, so turning your head or moving around the room won't break the connection.

If an optical outlet is not close by, radio channels are still available. Today, a portable phone provides services as long as we're close to civilization. With low-orbit communication satellites like Motorola Corporation's Iridium going up, cellular radio services will be available from anywhere on Earth. You'll have to leave your Companion at home to be out of touch.

Today, a radio or television tunes only one channel or frequency at a time because it has only one tuner circuit, whereas home VCRs allow you to record one channel while you watch another. Molecular technology will make it possible to build hundreds or thousands of tuners in a tiny space, all working simultaneously. Each tuner would have its own computer that selects and records broadcasts the user ready. Using this technology, the companion can extend the user's reach into information and entertainment channels tremendously.

Power

A variety of sources such as *photovoltaic cells*, *fuel cell*, *rechargeable batteries*, and *thermal-differential cells* provide power. A direct electrical connector will be provided for recharging if the batteries become completely drained. Several different power sources are necessary for two reasons: if one source—say, the solar batteries—isn't sufficient in a situation, such as darkness, other sources will take over. The other reason is that the total aggregate supply must be up to handling peak power drains on the system.

Power is the biggest limit to the Companion's capability. Radio transmission consumes the greatest amount of power, up to perhaps five watts, The display, earphones, and optical transmission each use less than 100 milliwatts for continuous duty. The computers and memory draw a negligible amount of power, less than one milliwatt. The Companion may be limited to a maximum of an hour or two of radio frequency broadcasting before automatic monitors prevent any further power drain. The power supply would have to deliver a maximum of about five watts, and the battery capacity needs to be at least 10 watt-hours.

Interface

The spoken word will be the cornerstone of the Companion's user *interface*. It will handle such casual requests as, "Is there anything worth watching tonight?" Continuous speech recognition, a long-desired goal of computer scientists, has seen some dramatic breakthroughs in recent years. Personal computers will increasingly incorporate this technology in the next few years, and common appliances are likely to incorporate voice recognition soon after.

Since speech isn't always appropriate or convenient, the Companion provides other *interfaces*. Eye-motion tracking sensors in the frames combined with projected visual control panels allow the user to operate the Companion through eye movement. This virtual display panel can also be operated by hand movement; built-in cameras locate the user's hands and fingers. When needed, realistic images of keyboards, levers, buttons, and knobs are projected into the user's field of view. The projected controls move and react the way real controls do. The controls do not provide force feedback, but and audible click or other sound effects make up for this shortcoming. Much of this technology is already available in the "*heads-up*" display helmets that fighter pilots use today, which eliminate the need to look down at a control panel while in flight.

Library

The Companion contains an extensive library of information and entertainment, including digital versions of books, music, movies, and more.

Databases

Books can be represented in a computer as graphic images of pages or as text, or both. By using images, the original look of the book is preserved, including typefaces and artwork. But images take up a lot of memory and are difficult, if not impossible, to search. Plain text is memory-efficient and searchable but a little dull. By using both techniques, millions of books can be stored.

In addition to books, the Companion's memory has enough room for many thousands of hours of video. Since many thousands of still art works fit in the same amount of memory as an hour of continuous motion video, large still art collection will be offered as well.

Personal empowerment is a key goal in the design of the Companion. To that end, the library will be well stocked with interactive computer software called courseware, which combines various media (narration, music, text, video, quiz sections) to educate the user in subjects ranging from first aid to French cooking. An owner's manual for the companion will also be included.

The companion will also be able to play many games. Old standbys--card games and chess—are offered along with twenty-first century video games and interactive electronic community events. Foot pedals, joysticks, and other equipment can be interfaced easily. Today, thousands of computer hobbyists across the country link up nightly to play shared games on computer networks. Games like these also will be possible with the Companion, but with far greater realism.

The library contains extensive and detailed maps. The maps combine with the computing and communications facilities to form a powerful navigation system. The *global positioning system* (GPS) a present-day satellite-based navigation system originally developed for military applications, allows a radio receiver to determine its location, accurate to a couple meters, anywhere on Earth. Thirty years from now GPS, or its eventual successor, will be able to locate any constructed object large enough to contain the required antenna—about ten centimeters. The companion can receive these signals to determine the user's location as well as the location of other users and specified objects.

The personal library contains audiovisual recordings the user makes, both live and copied from broadcasts. It also includes a personal medical history and a record of individual preferences in entertainment and news. Directories, notebooks and active calendars keep track of friends, relatives, acquaintances, business associates, important dates, and other changing information.

Knowledge Bases

The *Companion* contains knowledge bass for medicine, law, and finance. *Knowledge bases* are sets of facts, procedures, and rules of thumb that pertain to a given field. The medical knowledge base provides a multimedia resource covering anatomy, medicine, and psychology. It emphasizes self-help. And it ties in to the owner's private library to provide an ongoing personal medical history using data gathered from the biosensors. In case of an emergency, the Companion can automatically call for medical help.

The legal and financial knowledge gases provide the user with extensive advice on tackling business and life's sticky situations. Like the medial knowledge base, the financial and legal knowledge bases tie in to the private library that keeps track of the user's personal financial and legal data.

A survival knowledge base helps the owner cope with severe and life-threatening situations. It describes how to gather food, build shelter, treat injury, and deal with terrain and wildlife. Of course, with global communication capabilities, sending for help will be easy. Yet some people will want to challenge themselves by setting out deliberately with only the Companion's suggestions to show them how to survive. *Wilderness training* and *survival courses* enjoy popularity today because of the challenge and escape such activities provide. As technology progresses and life gets more complex, some people will crave the direct confrontation of physical survival more and more.

The library includes extensive langúage translation facilities and a cross-cultural etiquette knowledge base. The user can carry on a conversation in a foreign language and the Companion simultaneously translates a foreign tongue and presents it as displayed text or spoken words. *Automatic* or *machine language translation* is technically very challenging, but even if imperfect, machine translation can still be helpful in many situations.

The *etiquette knowledge* base contains information on cultural practices, protocols, and taboos. With this information, users can find out what's appropriate to bring to a birthday party in Warsaw, Ohio, or a wedding in Warsaw, Poland. Although you may never use the etiquette knowledge base in a practical situation, many people will find this information fascinating, and it will help us appreciate the way others live.

Additional Features

Security is provided at many levels. An identity file containing a retinal scan, blood chemistry profile, and speech patterns personalizes your Companion. This identity file is developed over time as you use your Companion and is continuously updated. It can be used to prevent another person from activating a stolen Companion. It can be used to prevent another person from activating a stolen Companion. In addition, you will be able to give explicit commands and passwords to make your personal information more private.

Computer viruses, "*Trojan Horses*," and other methods of hacking pose a greater problem than outright theft. Powerful computers in the larger communications system and in the Companion will probably be dedicated to watching for suspicious content in received data. Since some people want to protect information, and others want to get at it, data security is likely to remain a very tough issue regardless of the technological milieu.

One of the greatest advantages of having data computerized is that it can be searched rapidly. Techniques ranges from brute force methods that find every mention of the poet Robert Frost, to artificial intelligence searches that spot poetry in a Robert Frost style, and hypertext searches that let you know what Robert Frost's relatives did. Music and pictures can also be searched. At a minimum, descriptive text will accompany each movie and musical selection, which can be searched by commonly used text search methods. Neural networks and advanced *artificial intelligence* (AI) may be able to search the actual content of the selections for musical pharases, scenes, and the like. This would be like using the speed-search button on your VCR, if the VCR was smart enough to look for scenes you wanted to see.

With few exceptions, all this material is already available in computerized form. Maps, books and music have all undergone *computerization* in recent years. Video and movies will follow as the cost of memory continues to plummet. The main contribution nanotechnology makes to Companion is data storage big enough in bytes, yet tiny enough in volume, to carry all this material around with you, easily.

Alternatives

The hardware technology and packaging, interesting though it is, is only secondary in importance to the libraries and working software of the Companion. The physical form of the Companion, eyeglasses, is a convenient package because it gives easy access to the eyes and ears, is lightweight, is big enough to contain the necessary hardware, and is already an everyday sight.

But the Companion's technology may be delivered in other ways. One possible configuration would be similar to a pocket calculator with a foldout viewing screen, similar to Apple's Newton. Or the Companion could be designed as an implant, inserted into the body, powered by blood sugar, and connected directly to visual and auditory nerves. Assuming it will be harder to work out these biological details than to use known technologies for visual displays, audio, and so forth, an implant version of this technology—while inevitable—seems further in the future. One style alternative that may be quite attractive is wraparound goggles that would cover the entire visual field, producing an artificial-reality effect that is much more compelling.

Marketing and Costs

One of the promises of *molecular manufacturing* is to drastically reduce the costs of making things. In the Companion's case, the costs of design will probably be the greatest: organizing the data, designing the hardware, and writing the software will take a substantial effort. But the market is huge. Potentially everyone who has a phone or a

TV set will want one. So the costs of design and manufacture can be amortized over hundreds of millions of units. This would make the Companion so cheap that it could be sold for a very nominal amount, say, one dollar. The real cost to the end user would be determined by the services he or she chose to subscribe to.

Social Impact

So, where does all this fancy technology get us?

Media

To begin with, radio, telephone, and television will merge into a *single media resource*. The familiar paper print media of newspapers and magazines will be largely displaced; print reporters and editors will gradually move to this new medium. Social forms of information and entertainment (cinema, night clubs, town meetings) may remain but probably in an altered form.

For much of the history of broadcasting, the major networks were the only game in town; people had few choices in what they could see or hear. Recently, cable services and low-cost videotapes have made huge inroads in the networks' formerly sovereign territory. This trend toward entertainment on demand will most likely continue. Individual tastes will become more specialized, more esoteric. The Companion is well suited to cater to specialized tastes. Its huge storehouse of entertainment and its high-speed communication system allows entertainment consumers an enormous selection from which to choose. In addition, powerful search capabilities will be able to sort through a large selection quickly and automatically.

Politics

Because television broadcasting is still controlled by a few well-heeled interests, every society, from the most liberal to the most reactionary, has this medium manipulated to some degree by the government. The companion puts the technology of television broadcasting in everyone's hands. It may no longer be possible for large institutions to deliver an impenetrable wall of distorted information. If eyewitnesses are present, "*facts*" will be extremely difficult to control. Currently, personal communication, from letters to videotapes, are seen as politically subversive in some areas of the world. In such areas, devices as powerful as the Companion would be officially banned, though enforcing such a ban would be extremely difficult due to its cheapness, small size, and innocuous appearance.

If *social acceptance* is high, the Companion has the potential for becoming a new political tool. The Companion's software could include an electronic voting booth. Since the Companion will know the user's home address, it will present only those candidates and issues appropriate for his or her political district. Information and propaganda on the pros and cons of issues can be called up for display. The Companion sends its electronic ballots to a central voting system. Voting data could be encrypted to preserve vote secrecy. Vote counting would be very fast. Today, informal radio and TV polls are taken by people calling in and registering their opinions by Touch-Tone phone. These public opinion pools, while entertaining, carry little political weight. The use of the Companion in elections and referenda could change the face of politics forever.

Everyday Life

It's tempting to think that, as radio squelched the art of conversation, and television has assumed a central role in many of our lives, the Companion also will be divisive in personal relationships. Since it provides so much information at a personal level, it will undoubtedly help people become more independent. By catering to the most esoteric personal tastes, the Companion will make our already fragmented society even more so. However, although it's true that many people may become even more addicted to a never-ending stream of news and entertainment, remember that the Companion is a *two-way* communications device. It's designed to connect like-minded individuals and to aid social interaction with its etiquette and language libraries.

People will use their Companions to record their life's experiences as they do today with still and video cameras. It has such a voluminous memory that a person can, for example, let it record an entire two-week vacation, then review the results late rand (we can only hope) weed out the boring parts. One possible downside is that with all these cameras on all the time, people will be more circumspect about how they act in public, lest they be taught in some terribly embarrassing moment. "*Candid Camera*" and "*Funniest Home Videos*" may be just the tip of the iceberg. For this reason, it may become a social gaffe to wear your Companion to any gathering where mutual trust and privacy are important.

Because the Companion contains a highly personalized data bank that is tied in to communications networks, the potential exists for institutional abuse. Governments could use the Companion to keep track of political undesirables or foreign nationals. If the security were broken, a companion could function as a portable spy even without the owner's knowledge. *Demographics brokers* could secretly gather buying patterns and personal preferences to build targeted advertising lists. These unfortunate prospects of the device will lessen its acceptance by some. The Companion will have security features, but even so, many people will be reluctant to own one out of fear of having their privacy invaded.

All too often, advances in technology promise to make life easier but end up delivering only fleeting glamour and then endless complication. The Companion will be exciting, complicated, and will make life easier. Serving as a personal secretary, it will remind us of appointments and keep track of all those phone numbers we need to know. For the most part it will be unobtrusive, doing its own internal housekeeping, requiring no battery replacements, and bothering us only when we ask it to. The real challenge is deciding how to use such enormous informational resources. Undoubtedly, our values will shape how we use it, and using it will in turn shape our values.

Handicap Assistance

The wonderful possibilities of advanced molecular and biological technology give us the hope that we might eliminate physical handicaps. However, this will likely be very difficult to achieve; many of us may be blind and deaf for some time to come. Nevertheless, the Companion, while do not warm and fuzzy like a seeing-eye dog, can be of service to the handicapped.

The *blind can* benefit in several ways. Neural network and pattern recognition technology may be sufficiently advanced to program everyday objects into a visual memory and to allow an ongoing learning process for unfamiliar objects. The Companion's video cameras, surveying the scene, can provide a running verbal narrative about the scene, can provide a running verbal narrative about the scene before you. Likewise, a hearing-impaired person could "*listen*" to someone talking. The Companion would pick up the audio and translate it in real-time into a text or sign-language window displayed in his visual field, like closed-captioned television today. Discriminating the right sounds may be a problem though: picking out one voice in a crowded or noisy room may be as difficult for the Companion as it is for a person.

People with a physical handicap that limits their movement could use the Companion as a control center. Devices such as appliances, doors, robot arms, and the like can be designed to be controlled by the Companion. Using eye movement to drive these interfaced devices, a person could run a household of automated equipment even if he or she were completely immobilized. If biotechnology doesn't solve the problems of aging, the Companion can also help the elderly. They can either summon help at a moment's notice or have the Companion summon help automatically, sending a signal if the user's vital signs look bad. The Companion could also aid their hearing and assist visually, if needed.

Normally healthy people can have their senses enhanced as well. The cameras and microphones both cover a broader range of light and sound than our natural senses can detect. The Companion can be used with infrared or low-light capabilities to find your way around in the dark, or with audio sensitivity boosted to listen to faint sounds.

In combination with the appropriate software, the sensory apparatus can do much more. The *microphones* can detect a specific sound from a jumble of noise. The cameras can pick out an object and zoom in on it. This would be an obvious boon to bird-watchers, since the bird's call can be automatically isolated and identified. In addition, it could help us all of those terribly awkward moments when someone recognizes us and calls us by name, but we've forgotten the other's name.

Barriers and Breakthroughs

Technical

The Companion depends heavily on *nanotechnology*, with the present level of technical activity, it seems likely that nanotechnology breakthroughs such as assembler construction may take place as early as the year 2010. The Companion's display, communication, power supplies, and computers are all refinements of currently available technologies. The only effort would be to translate these known technologies to a new form, similar to designing a radio using transistors instead of vacuum tubes.

The Companion assumes only moderate advances in current communications technology; it requires no major breakthroughs. Multi channel radio frequency transmission has been in practice for the better part of a century. Fiber optic communications development is well underway and will continue to replace copper wire as the bulk communications medium of choice; we can expect to see fiber optic connections in some homes by the turn of the century.

Social

With television and radio slowly migrating to cable and other media, slices of the airwaves, or radio-frequency spectrum, will be freed up for other uses. The process of deal locating and reallocating the airwaves isn't a free-for-all; it's heavily government regulated and requires years to hammer out each detail. The Companion would have go through approval by the FCC and the analogous agencies of other governments before necessary radio bands can be allocated.

Currently, the most difficult barrier to producing the Companion seems to be our unresolved struggle over intellectual property rights. Digitized books, movies, and music must be created somewhere. Much of this material is currently in the public domain, and more will be by the time the Companion is developed. But under existing *copyright* law, newly created works will be protected for as long as seventy years. Copyright holders will naturally want compensation. The legal and economic process of putting all this information and entertainment into the public's hands will be quite complicated.

If we are unable to establish a technical means for rewarding the holders of existing copyrights, the Companion can still be sold with a standard library of public domain works, and the user would have the option of purchasing copyrighted works from various vendors. The user, of course, won't have to decide at the time of manufacture: data packages can be transferred into the Companion easily at any time over the radio or optical channels.

Pay-per-use is another alternative. A huge library could be delivered including all the desirable movies, music, and games, but some material could be locked out. You would call an entertainment service that unlocks the requested items and bills your account. This is now done with *pay-per-view* television and software delivered on optical disk. In addition to copyright complications, legal and medical professionals, fearing an impact on their careers, may resist providing cheap availability of knowledge bases gathered from their fields. This fear may be unfounded; many self-help books on medicine and law already provide help for individuals without adversely affecting these professions.

2020 Vision

The most formidable breakthrough needed to produce the Companion will be the development of nanotechnology itself. Computerized libraries of information and entertainment will emerge with or without nanotechnology, and the process is already well underway. Based on these trends, we can project the Companion to arrive on the scene around the year 2020.

15

NANOBOT CONSTRUCTION

Molecular manufacturing—the heady motion of assembling almost anything, from computers to caviar, from individual molecules—would change the world, if someone could just find a way to make it work. Imagine nanoassemblers: busy work gangs of "*pick and place*" robotic manipulator arms, each one tens of nanometers in size. Con trolled from on high by a powerful computer, these simple devices would arrange blocks of molecules to make copies of themselves—and these machines would, in turn, build still other nanomachines, which, in turn, would create others, and so on in an exponential expansion. These nanobot construction crews could then be directed to accomplish astonishing tasks such as curing diseases from inside the body and fabricating intricately engineered materials from basic bulk feedstocks at extraordinarily low cost.

For years, futurist K. Eric Drexler and his colleague Ralph C. Merkle, the noted *cryptographer*, nurtured this vision, using computer simulations of nanometer-scale gears, pumps and other molecular machine subsystems. In these images, coloured spheres indicate the position of each component atom. Several analytical critiques of these elaborate simulations suggested that the devices simply wouldn't work as hoped. But perhaps the most biting criticism focused on their status as mere digital representations and not real-world objects interacting with complex chemical bonding forces in the nanoscale environment, where macroscale physics doesn't always apply and quantum mechanics often prevails.

Then, in 1999, Merkle decided to leave a long-term position at the Xerox Palo Alto Research Center to try to give real sub stance to his computer-concocted concepts. He joined soft ware mogul James R. Von Ehr II, an admirer of Drexler's ideas, who had decided to spend part of the "*low-nine-figure*" fortune he garnered from the sale of his company, Altsys, to start the first "*molecular nanotechnology*" company.

Zyvex, the Richardson, Texas-based company Von Ehr founded, is aiming high. "We'd love to the Applied Materials for the manufacturing base of the world," he says, referring to the world's leading semiconductor equipment manufacturer. But his molecular nanotechnology company is still in the early stages of its quest. Its experience has already demonstrated the difficulty of moving from software-wrought molecular machines to a real-world nanoassembler. One of the first things the company had to do was to pull hack from the idea of building things atom by atom. Today the long-term focus is on *fabricating nanoscale machine systems* from large molecules or blocks of molecules. To that end, company scientists are experimenting with the bottom-up approach to making assemblers with molecular nanotechnology—learning to put together structures molecule by molecule using atomic force microscopes and the like.

In the new term, Zyvex researchers are developing *microelectromechanical systems* (MEMS), whose structures mea sure in tens of microns. MEMS devices are manufactured using a top-down approach. They are lithographically patterned and etched out from silicon substrates or other materials.

Richard Feynman, the ur-guru of the field, did say that bigger machines could be used to make smaller machines. And that's where MEMS comes in. According to Merkie, MEMS fabrication methods can be used to build robotic assembly arms. These microelectromechanical "*hands*" can assemble submicron hands that can build even tinier hands, and so forth, step by step until the ultimate in mechanical smallness is achieved: the nanoassembler, a nanometer-scale manufacturing device. For Zyvex to accomplish its goal, the MEMS robots that it creates must be shrunk by a factor of 1,000 or so. Once the requisite down-sizing has been completed, nanobots, as envisaged, will be used for "*atomically precise manufacturing*" to make virtually anything. Universal constructors could manufacture a Rolex watch, followed by a computer memory and then by nanomachines that can treat diseases.

Two basic ideas underlie Zyvex's *road to nanorobotics*, Merkle says. In position.' assembly, mechanical manipulators pick up and precisely place objects into assemblies. Accurate positional control at scales of tens of nanometers means moving large molecules or blocks of molecules where you want them and causing them to bond to an assembly as desired.

A machine that can build a complex material or device from the bottom up—using an *Erector set of molecules* or *chunks of molecules* for components—will require one other critical technology: the machine must be able to make a copy of itself. "If you wish to achieve economical production of molecular devices in large numbers, some form of self-replication is necessary," Merkie explains. And unless you had hordes of nanoconstructors on the job, building macroscale objects molecule by molecule would take a very long time. Zyvex researchers note that a fully self-replicating machine would not be necessary to manufacture an assembler. Still, the task is a monumental one: any machine outside the biological sphere that could make any number of copies of itself would probably lead to a Nobel Prize, possibly several.

So how does Zyvex get to an assembler? Recently Zyvex joined with Standard MEMS in Burlington, Massachusetts—a company that fabricates microsystems—in a two-year collaborative program to develop lithographically formed micron-scale manipulator arms and grippers that can assemble smaller manufacturing devices from pallets of precisely positioned "*active*" parts produced on silicon wafers. "If the positional accuracy is good enough, the arm simply has to reach down and pick the part up. You don't need an elaborate sensing system," Merkle says.

According to Von Ehr, Zyvex researchers have developed a method by which the MEMS manipulator can detach the parts from the substrate so they are ready for microassembly using self-centering *snap-connectors* a useful, if not exactly new, capability. Manipulator parts would be created *lithographically*, etched out and then disconnected from the surface substrate so they could he picked up by the MEMS manipulator and attached to the device by snapping them into place. Von Ehr hopes these MEMS manipulators will prove to be an intermediate technology that could serve as a moneymaking product, such as a device to align a fiber-optic cable, as the company pursues its goals.

Unfortunately, it's hard to find anyone in the MEMS field who would be interested in this kind of technology or anyone who can contemplate what kind of product it would actually be used to build, says Kaigham J. Gabriel, now professor of electrical and computer engineering and robotics at Carnegie Mellon University and former director of the MEMS program at the Defense Advanced Research Projects Agency.

And what about the *nanoscale assembler*? "There is significant controversy about how long it will take to achieve ultimate control at the molecular level," Merkie notes, adding that "it could take a decade or two."

So Zyvex has its work cut out for it. But it has embarked on a mission to achieve the nano equivalent of a moon shot in relative isolation with a staff of only 37. In general, the scientific research community has distanced itself from this project. Several scientists working in the field of nanotechnology derided Zyvex's scheme but requested anonymity to avoid protests from amateur nanotech enthusiasts. Says one researcher, "We've seen no experimental proof that any portion of their scheme can actually be accomplished. We think it's a lot of nonsense." Merkle responds that "nothing we propose contradicts the laws of physics."

Meanwhile, with Von Ehr's fortune backing up the research effort, Zyvex can continue to function for many years without having to market a product, but the entrepreneur allows that it would be nice to make some money. "This whole thing is a lot harder than it first seemed," he admits. Von Ehr has reportedly already spent about $20 million on the project, hut considering last year's stock market Fall and the reality of the task looming on the horizon, he says he is planning to seek outside investment once financial conditions improve. "If we're going to grow," he says, "we'll need more money." No matter how big the envisioned payoff, however, given the daunting technical difficulties and a 10-to 20-year timeline to possible success, one wonders just who would be willing to put up significant investment funds. Perhaps Zyvex's trek toward molecular nanotech could be financed by small contributions from its legions of true believers.

16

MINIATURIZATION

Do we really need to keep on making circuits smaller? The *miniaturization* of silicon microelectronics seems so inexorable that the question seldom comes up—except maybe when we buy a new computer, only to find that it becomes obsolete by the time we leave the store. A state-of-the-art microprocessor today has more than 40 million transistors; by 2015 it could have nearly five billion. Yet within the next two decades this dramatic march forward will run up against scientific, technical and economic limits. A first reaction might be, So what? Aren't five billion transistors enough already?

Yet when actually confronted with those limits, people will no doubt want to go beyond them. Those of us who work to keep computer power growing are motivated in part by the sheer challenge of discovering and conquering unknown territory. But we also see the potential for a revolution in medicine and so many other fields, as extreme miniaturization enables people and machines to interact in ways that are not possible with existing technology.

As the word suggests, *microelectronics* involves components that measure roughly one micron on a side (although lately the components have shrunk to a size of almost 100 nanometers). Going beyond microelectronics means more than simply shrinking components by a factor of 10 to 1,000. It also involves a paradigm shift for how we think about putting everything together.

Microelectronics and *nanoelectronics* both entail three levels of organization. The basic building block is usually the *transistor* or its *nano equivalent*—a *switch* that can turn an electric current on or off as well as amplify signals. In microelectronics, transistors are made out of chunks of semiconductor—a material, such as impure silicon, that can be manipulated to flip between conducting and nonconducting states. In nanoelectronics, transistors might be organic molecules or nanoscale inorganic structures.

Interconnection

The next level of organization is the *interconnection*—the wires that link transistors together in order to perform arithmetic or logical operations. In microelectronics, wires are metal lines typically hundreds of nanometers to tens of microns in width deposited onto the silicon; in nanoelectronics, they are *nanotubes* or other wires as narrow as one nanometer.

At the top level is what engineers call *architecture*—the over all way the transistors are interconnected, so that the circuit can plug into a computer or other system and operate independently of the lower-level details. *Nanoelectronics researchers* have not quite gotten to the point of testing different architectures, but we do know what abilities they will be able to exploit and what weaknesses they will need to compensate for.

In other ways, however, microelectronics and nano electronics could not be more different. To go from one to the other, many believe, will require a shift from top-down manufacturing to a bottom-up approach. To build a silicon chip today fabrication plants start with a silicon crystal, lay down a pattern using a photographic technique known as lithography, and etch away the unwanted material using acid or plasma. That procedure simply doesn't have the precision for devices that are mere nanometers in width. Instead researchers use the methods of synthetic chemistry to produce building blocks by the mole (6×10^{23} pieces) and assemble a portion of them into progressively larger structures. So far the progress has been impressive. But if this research is a climb up Mount Everest, we have barely just reached the base camp.

The use of molecules for electronic devices was suggested more than a quarter of a century ago in a seminal paper by Avi Aviram of IBM and Mark A. Ratner of Northwestern University By tailoring the atomic structures of organic molecules, they proposed, it should be possible to concoct a transistor like device. But their ideas remained largely theoretical until a recent confluence of advances in chemistry, physics and engineering.

Of all the groups that have turned Aviram and Ratner's ideas into reality, two teams—one at the University of California at Los Angeles and Hewlett-Packard, the other at Yale, Rice and Pennsylvania State universities—stand out. Within the past year, both have demonstrated that thousands of molecules clustered together can carry electrons from one metal electrode to another. Each molecule is about 0.5 nanometer wide and one or more nanometers long. Both groups have shown that the clusters can behave as on/off switches and might thus be usable in computer memory; once on, they will stay on for 10 minutes or so. That may not sound like a long time, but computer memory typically loses its information instantly when the power is turned off; even when the power is on, the stored information leaks away and must be "*refreshed*" every 0.1 second or so.

Although the details differ, the switching mechanism for both molecules is believed to involve a well-understood chemical reaction, oxidation reduction, in which electrons shuffle among atoms within the molecule. The reaction puts a twist in the molecule, blocking electrons as surely as a kink in a house blocks water. In the "*on*" position, the clusters of molecules may conduct electricity as much as 1,000 times better than in the "*off*" position. That ratio is actually rather low compared with that of typical semiconductor transistors, whose conductivity varies a million fold. Researchers are now looking for other molecules with even better switching properties and are also working to understand the switching process itself.

The research group at Harvard University is one of several that have focused not on organic molecules but on long, thin inorganic wires. The best-known example is the *carbon nanotube*, which is typically about 1.4 nanometers in diameter. Not only can these nanoscale wires carry much more current, atom for atom, than ordinary metal wires, they can also act as tiny transistors. By functioning both as interconnections and as components, *nanowires* kill two birds with one stone. Another advantage is that they can exploit the same basic physics as standard silicon microelectronics, which makes them easier to understand and manipulate.

In 1997 Cees Dekker's group at the Delft University of Technology in the Netherlands and Paul L. McEuen's group, then at the University of California at Berkeley, independently reported highly sensitive transistors made from metallic *carbon nanotubes*. These devices could be turned on and off by a single electron but required very low temperatures to operate. This past July Dekker's team swept away this limitation. The researchers used an atomic force microscope to create a single-electron transistor that could function at room temperature. Dekker and his co-workers have also fashioned a more conventional field-effect transistor, the building block of most integrated circuits today, out of a *carbon nanotube*, and McEuen's group has combined metallic and semiconductor nanotubes into a diode, which allows electric current to pass in one direction only Finally my group has demonstrated a very different type of switch, a nanoscale electromechanical relay.

A major problem with nanotubes is that they are difficult to make uniform. Because a slight variation in diameter can spell the difference between a conductor and a semiconductor, a large batch of nanotubes may contain only a few working devices. In April of this year Phaedon Avouris and his colleagues at the IBM Thomas J. Watson Research Center came up with a solution. They started with a mixture of *conducting* and *semiconducting nanotubes* and, by applying a current between metal electrodes, selectively burned away the con ducting ones until just semiconducting ones were left. The solution is only partial, however, because it requires the use of conventional lithography to wire up the random nanotube array and then test and modify each of the individual elements, which would ultimately number in the billions.

This group has also been working on a different type of nanoscale wire, which then term the *semiconductor nanowire*. It is about the same size as a carbon nanotube, but its composition is easier to control precisely. To synthesize these wires, we start with a metal catalyst, which defines the diameter of the growing wire and serves as the site where molecules of the desired material tend to collect. As the nanowires grow, we incorporate chemical dopants (impurities that add or remove electrons), thereby controlling whether the nanowires are n-type (having extra electrons) or p-type (having a shortage of electrons or, equivalently, a surfeit of positively charged "*holes*").

The availability of *n-* and *p-type* materials, which are the essential ingredients of transistors, diodes and other electronic devices, has opened up a new world for us. We have assembled a wide range of devices, including both major types of transistors (field-effect and bipolar); inverters, which transform a "0" signal to a "1"; and light-emitting diodes, which pave the way for optical interconnections. Our bipolar transistors were the *first molecular-scale* devices ever to amplify a current.

Building up an *arsenal* of molecular and nanoscale devices is just the first step. Interconnecting and integrating these devices is perhaps the much greater challenge. First, the *nanodevices* must be connected to molecular-scale wires. To date, organic-molecule devices have been hooked up to conventional metal wires created by *lithography*. It will not be easy to substitute nanowires, because we do not know how to make a good electrical connection without ruining these tiny wires in the process. Using nanowires and nanotubes both for the devices and for the interconnections would solve that problem.

Second, once the components are attached to nanowires, the wires themselves must he organized into, for example, a two-dimensional array. In a report published earlier this year, Duan and another member of the team, Yu Huang, made a very significant breakthrough: they assembled *nanocircuits* by means of fluid flows. Just as sticks and logs can flow down a river, *nanoscale wires* can be drawn into parallel lines using flu ids. In the lab scientists have used ethanol and other solutions and controlled the liquid flow by passing it through channels molded into polymer blocks, which can be easily placed on the substrate where they wish to assemble devices.

The process creates interconnections in the direction of the fluid flow: if the flow is along only one channel, then parallel nanowires are formed. To add wires in other directions, the flow is restrict and repeat the process, building up additional layers of nanowires. For instance, to produce a right-angle grid, we first lay down a series of parallel nanowires, then rotate the direction of flow by 90 degrees and lay down another series.

Our approach which is similar to that being pursued by the team at U.C.L.A. and Hewlett-Packard, is deterministic. We are trying to create arrays with a certain predictable behaviour. Form follows function. An alternative proposed by the group at Rice, Yale and Penn State is to allow blocks of devices and wires to interconnect at random. Later, the ensemble can be analyzed to determine how it might be used for storage or computation. In this case, function follows form. The problem with this procedure is that it would take a huge effort to map a complex network and figure out what use it could be put to.

Intimately linked to all these efforts is the development of architectures that best exploit the unique features of nanoscale devices and the capabilities of bottom-up assembly. Although we can make *unfathomable numbers* of *dirt-cheap nanostructures*, the devices are much less reliable than their *microelectronic counterparts*, and our capacity for assembly and organization is still quite primitive.

In collaboration with Andre DeHon of the California Institute of Technology, the team has been working on highly simplified architectures that can be generalized for universal computing machines. For memory, the architecture starts with a two-dimensional array of crossed nanowires or suspended electromechanical switches in which one can store information at each cross point. The same basic architecture is being pursued by researchers at U.C.L.A. and Hewlett-Packard, and it resembles the magnetic-core memory that was common in computers of the 1950s and 1960s.

To overcome the unreliability of individual nanodevices, we may rely on sheer numbers—the gizmos are so cheap that plenty of spares are always available. Researchers who work on defect tolerance have shown that computing is possible even if many of the components fail, although identifying and mapping the defects can be slow and time-consuming. Ultimately we hope to partition the enormous arrays into subarrays whose reliability can be easily monitored. The optimum size of these subarrays will depend on the defect levels typically present in molecular and nanoscale devices.

Another significant hurdle faced by *nano electronics* is "*strapping.*" How do engineers get the circuit to do what they want it to? In *microelectronics*, circuit designers work like architects: they prepare a blueprint of a circuit, and a fabrication plant builds it. In nanoelectronics, designers will have to work like computer programmers. A *fabrication plant* will create a raw nanocircuit—billions on billions of devices and wires whose functioning is rather limited. From the outside, it will look like a lump of material with a handful of wires sticking out. Using those few wires, engineers will somehow have to configure those billions of devices. Challenges such as this are what keeps the scientists tremendously excited about the field as a whole.

Even before these problems are solved, *nanodevices* may have useful applications. For example, *semiconducting carbon nanotubes* have been used by Hongjie Dai's group at Stanford University to detect gas molecules, and Yi Cui has used semiconductor nanowires as *ultrasensitive detectors* for a wide range of biological compounds. Scientists have converted nanowire field-effect transistors into sensors by modifying their surfaces with molecular receptors. This technology has the potential of detecting single molecules using only a voltmeter from a hardware store. The small size and sensitivity of nanowires also make possible the assembly of extremely powerful sensors that could, for instance, sequence the entire human genome on a single chip and serve in minimally invasive medical devices. In the nearer term, we could see hybrids of micro and nano: silicon with a *nano core*—perhaps a high-density computer memory that retains its contents forever.

Although substantial work remains before nanoelectronics makes its way into computers, this goal now seems less hazy than it was even a year ago. As we gain confidence, we will learn not just to shrink digital microelectronics but also to go where no digital circuit has gone before. *Nanoscale devices* that exhibit quantum phenomena, for example, could be exploited in quantum encryption and quantum computing. The richness of the nanoworld will change the macroworld.

INDEX

A
Ab initio, 152
Active, 267
Adenosine triphosphate, 174
Adrenalin, 194
Aerodynamics, 244
African quagga, 181
Aging is natural, 176
Agoraphobia, 97
Aircar, 240, 244
Aircraft, 246
Airtrucks, 248
Akamba, 3
Akathisia, 96
Alcmaeon of Croton, 11
Alcoholism, 95
Alzheimer's disease, 96
Ambroise pare, 19
Ameliorate, 205
American Physical Society, 100
Amoeba proteus, 73
Amorphous films, 139
Anaphylaxis, 95
Ancestors, 220
Androphobia, 97
Angioedema, 95
Anthracosis, 8
Antibiotics, 41, 202
Antiseptic, 34
Apocalyptic nightmares, 99
Apocryphal queen, 224
Apollo, 227
Apollo Project, 230
Arachnophobia, 97
Architecture, 270
Aristotelian, 14
Aristotle, 14, 115
Arsenal, 272
Artificial, 154
Artificial cells, 81
Artificial intelligence, 125, 221, 259
Asclepiades, 15
Asclepius, 10
Associative computers, 122
Asteroidal winter, 182
Astounding Science Fiction, 78
Astrology, 94
Atomic force microscope, 100
Atomic layer epitaxy, 142
Atomically precise manufacturing, 266
Attractions, 195
Autogenous creatures, 213

Autogenous molecular manufacturing, 240
Autogenous technology, 217, 222
Automated engineering, 183
Automatic, 258
Automotive technology, 204

B

Bacteria, 158
Barbiturates, 41
Beanstalk, 236
Bell System Technical Journal, 112
Belonephobia, 97
Beneficent, 155
Benzodiazepines, 41
Billions, 182
Biosphere, 200
Biostasis, 187
Biotechnology, 156
Black Death, 212
Black-goo, 106
Blades, 241
Blood sports, 195
Bottom line, 115
Bottom-up fabrication, 102
Bottom-up methods, 172
Break junction, 171
Brownian motion, 115
Bulimia, 95

C

California Institute of Technology, 100
Candid Camera, 261
Cannonball, 237
Carbon nanotube, 271, 272
Catastrophic risk, 151
Catastrophic risks, 150, 152, 159, 161
Cecil Textbook of Medicine, 43
Cell repair machines, 175, 176
Cell repair technology, 181
Channel, 113
Charles Turner Thackrah, 22
Cheating, 224
Chimeras, 150, 155
Chipping stones, 214
Chirurgia, 13
Chlorpromazine, 41
Chocoholics, 94
Christian, 17
Chromophobia, 97
Chunks of molecules, 267
Classical kind, 136
Clocks, 178
Coffee, 94
Cola drinkers, 94
Command economy, 220
Communication, 128
Companion, 251, 257
Computer, 197, 253
Computer chip, 131, 165
Computerization, 259
Conducting, 272
Consumer synthesizers, 217
Copy, 192
Corpus, 12
Cro-Magnons, 2
Crossing, 154
Crusoe, 222
Cryonics community, 104
Cryptographer, 265
Cultural stagnation, 184

D

Data storage capacity, 254
Daughter genes, 154
Decoherence, 124
Demographics brokers, 262
Deoxyribose, 121
Destructive nanomachines, 106
Development Principles, 109
Diamond, 119

Dinosaur, 182, 202
Dip-pen lithography, 170
Dipsomania, 95
Directed molecular evolution, 154
Dirt-cheap nanostructures, 273
Disease idealism, 49
Disney Corporation, 195
DNA fragments, 181
DNA shuffling, 154
Domination, 94
Downside risks, 158
Dubious risk, 158
E
E. coli, 53, 154
Earpieces, 253
Earth's environment, 233
Ecosystem protectors, 192
Electron-beam instruments, 166
Electronic components, 164
Electronic nanocomputers, 118
Electronic properties, 140
Electronics, 125
Electronics industry, 166
Electrons tunneling, 147
Electrostatic, 117
Elephants, 237
Eminent roboticists, 221
Empedocles, 11
Engines of Creation, 69, 81, 158, 232
Entropy, 111
Environmentalism, 227
Epitaxial growth, 139
Erasistratus, 14
Erector set of molecules, 267
Ethical problems, 107
Etiquette knowledge, 258
Evipan, 41
Ex vivo, 5
Extinct, 182
Extinction, 181
Eyeglass frames, 253
F
Fabricating nanoscale machine systems, 266
Fabricating nanostructures, 167
Fabrication plant, 274
Facts, 260
Feels like nothing, 232
Femtotechnology, 116
Ferrite core, 216
Fiber optic outlets, 254
First molecular-scale, 272
Fistula clubs, 22
Fluffing, 133
Flying saucer, 244
Folliculalgia, 95
Force, 134
Foresight Guidelines on Molecular Nanotechnology, 109
Fossil fuels, 204
Friedrich Engels, 22
Fuel cell, 255
Full-sensory telepresence, 250
Fun, 195
Funniest Home Videos, 261
G
Galen, 16
Galen of Pergamum, 15
Galenic medicine, 17
Garden protectors, 193
Genetic engineering, 132, 213
Genomics, 152
Geosynchronous orbit, 237
Giant superintelligent computer, 220
Global positioning system, 257
Gluttony, 94
Goose-feather pen, 170
Gravidophilia, 94
Gray goo problem, 158

Gray-goo, 105
Green revolution, 230
Ground traffic controller, 245
H
Hands, 266
Hans Moravec, 221
Hardware associative, 122
Heads-up, 256
Helicobacter pylori, 96
Helicopter, 241
Helmet, 234
Hemophobia, 97
Herophilus, 14
Hertfordshire, 20
Hippocrates, 11
Hippocratic, 17
Hippocratic Corpus, 11
Hippocratic medicine, 11
Hit, 112
Hobbies, 189
Holes, 272
Holism, 58
Home crafts, 193
Homo epitaxy, 140
Homo erectus, 211, 214
Homo sapiens, 2, 200
Human figurines, 190
Humanoid telerobot, 199
Humors, 12
Hypochondria, 94
I
Immune to changing times, 194
Imposing information, 223
In cyto, 52, 57
In situ, 78
In vitro, 43
In vivo, 5, 52, 55, 76, 86, 88, 92
Individual nanomachines, 132
Inked, 167
Inks, 170
Innocent, 224
Intelligence amplification, 126
Interconnection, 270
Interface, 255, 256
Interleukin 4, 155
Ironic, 115
Irreversibility, 114
J
Jewish, 17
K
Karl Marx's, 22
King Asa, 17
Kleptomania, 94
Knowledge bases, 257
Knowledge policy, 150
Knowledge worker, 125
L
Library, 120
Librium, 41
Limits to Growth, 227, 230
Liquid, 133
Liquid-fuel rocket engine, 236
Listen, 262
Listeria monocytogenes, 88
Lithium carbonate, 41
Lithographically, 267
Lithography, 273
Low-nine-figure, 266
Lunar soil, 228
Lush garden communities, 228
M
Machine language translation, 258
Macroscale, 126
Magically, 113
Magnetic thin films, 145
Malevolent, 155
Map, 100
Martial, 15
Masai, 3

Mask, 165
Master, 165
Mechanical nanocomputers, 119
Medical nanomachines, 103
Medieval weapons, 194
Memory chips, 164
Mercury capsule, 246
Methodology, 99
Micro-electromechanical systems, 266
Microchip, 165
Microcontact printing, 167
Microelectronic counterparts, 273
Microelectronics, 164, 269, 274
Microfluidic chips, 167
Microgravity, 231
Micromolding, 168
Micron sphere, 127
Micron-scale machines, 134
Microphones, 263
Microprocessors, 122, 164
Mind Children, 211
Miniature electronics, 251
Miniaturization, 102, 269
Minimal genome, 154
Mirrors, 228
Mobilophilia, 94
Modern jetliners, 246
Molecular dissasemblers, 249
Molecular engineering, 194
Molecular engineers, 176
Molecular intelligence, 193
Molecular manufacturing, 259, 265
Molecular manufacturing techniques, 103
Molecular memory, 128
Molecular nanotechnology, 266
Molecular switches, 118
Moller, 240
Money, 220
Moore's Law, 197
Moral risks, 150, 152
Multiprocessor, 124
Multiprocessor computer, 128
Mutually assured destruction, 152
N
Nano core, 274
Nano electronics, 274
Nano equivalent, 269
Nanoassociative memory, 122
Nanocameras, 105
Nanocircuits, 273
Nanocomputer, 117, 121, 122
Nanodevices, 272, 274
Nanodimension, 102
Nanoelectronic memory, 122
Nanoelectronics, 103, 117, 118, 121, 269
Nanoelectronics researchers, 270
Nanoengineered knife, 134
Nanoengineered wings, 242
Nanoengineering, 241
Nanofabrication, 164, 166
Nanoimprint lithography, 169
Nanomachine, 115, 118, 128, 129, 133, 173, 187, 199, 231
Nanomedicine, 4, 5, 81, 201, 202, 213, 226, 230, 231
Nanometer-scale components, 131
Nanometers, 119
Nanomicrophones, 105
Nanoparticles, 105
Nanorobots, 121
Nanoscale, 105, 117, 126
Nanoscale assembler, 268
Nanoscale devices, 275
Nanoscale wires, 273
Nanoscience, 158
Nanosecond, 118
Nanosensors, 105
Nanostructures, 164, 165, 173

Nanotech applications, 205, 247
Nanotech data link, 249
Nanotech defenses, 204
Nanotech era, 246
Nanotech recycling, 201
Nanotech-based solar power satellites, 238
Nanotechnology, 99, 101, 125, 156, 158, 174, 183, 189, 191, 196, 197, 200, 204, 206, 207, 212, 216, 226, 248, 252, 253, 263
Nanotechnology culture, 193
Nanotubes, 270
Nanoweaponry, 105
Nanowires, 271
Napoleon, 24
National Science Foundation, 215
Natural balance, 203
Nature, 154, 155
Nature's nanomachines, 176
Neanderthal, 2
Nerva, 203
New Scientist, 151
Noctiphobia, 97
Non plus ultra, 104
Normative analysis, 99
Nuclear winter, 151
Nymphomania, 94

O

On Medicine, 16
O'Neill colonies, 228
O'Neill-style colony, 231
Ontological, 162
Ophidiophobia, 97
Optical communications, 129
Orion, 203
Orthodoxy, 227
Osteoporosis, 230

P

Paper, 168
Parachute, 241
Parthenophobia, 97
Pathological liars, 94
Pay-per-use, 264
Pay-per-view, 264
Peer-to-peer communication, 129
Per second, 165
Percivall Pott, 22
Personal deterioration, 185
Personal empowerment, 256
Pharmacologia, 61
Phased array optics, 252
Phengophobia, 97
Phobophobia, 97
Photocells, 130
Photoelectric effect, 116
Photolithographic tools, 165
Photolithography, 102, 165
Photoresist, 165
Photovoltaic cells, 255
Physical computer, 198
Pick and place, 265
Picosecond, 122
Picotechnology, 116
Pier, 237
Pilot error, 244
Plague, 18
Planet-healing machines, 180
Play, 195
Pleiotropism, 152
Political field, 150
Polycrystalline films, 139
Polydimethylsiboxane, 167
Polygamy, 94
Praxagoras, 14
Prisoner's Dilemma, 224
Problem, 158
Problem nanotechnology, 225
Prognosticator, 225
Prophylaxis, 58
Protective, 158
Pulmonary silicosis, 8

Q

Qualitatively, 176
Quantum computer, 124
Quantum computing, 123
Quantum Dot Corporation, 172
Quantum dots, 149, 172
Quantum mechanics, 123
Quantum nanocomputers, 124
Quantum wells, 142
Quantum wires, 149
Qubits, 124

R

Rabbits, 237
Railroads, 248
Raise the stakes, 158
Rechargeable batteries, 255
Recombinant DNA techniques, 152
Refreshed, 271
Reincarnation, 211
Related risks, 155
Remote, 158
Repair machines, 187
Replicating nanomachine, 203
Replicators, 132
Reproductive physiologist, 182
Research replicators, 218
Restricting birthrates, 201
Reversibility, 122
Richard Wiseman, 20
Rides, 195
Road to nanorobotics, 266
Robotics, 80, 221
Robots, 222
Rockets, 235, 247
Rome, 15
Runner's high, 94

S

Salmonella typhimurium, 76
SAM, 168
Satyriasis, 94
Scanning tunneling microscope, 119, 169
Scanning tunneling microscopy, 100
Schistosomiasis, 8
Schizophrenia, 95
Scotophobia, 97
Script kiddies, 203
Second law of thermodynamics, 113
Security, 258
Self monolayer, 168
Self-consciousness, 153
Semiautonomous, 206
Semiconducting carbon nanotubes, 274
Semiconducting nanotubes, 272
Semiconductor, 147
Semiconductor nanowire, 272
Sentimental value, 194
Shake-and-bake, 140
Shape, 133
Shuttle flight, 207
Siderophobia, 97
Simplified, 160
Single media resource, 260
Single stage to orbit, 234
Skin drag, 242
Skyhook, 236
Smokers, 94
Snap turns, 195
Snap-connector, 267
Social acceptance, 260
Soft lithography, 167
Solid rocket boosters, 234
Space elevator, 236
Space shuttle, 234, 246
Specific Design Guidelines, 110
Spendoholics, 94
Spheres, 127
Stamp out, 132
Staphylococcus epidermidis, 53

Start wars, 185
Step-and-flash, 169
Stereo components, 129
Stone Age, 5
Stored in, 114
Strapping, 274
Stratosphere, 246
Streptococcus, 48
Stuff, 136
Sulphonamides, 41
Sun Microsystems, 105
Superconductor, 140
Surface nanomachines, 130
Survival courses, 258
Sweet tooth, 94
Switch, 118, 269
Synthesis techniques, 136
Synthesizer, 222
Synthesizer sitting, 216
Synthetic, 154

T

The Apologie and Treatise, 20
The Epic of Gilgamesh, 186
The Future and Its Enemies, 93
The Society of Mind, 66
The Temperature in Diseases, 25
Thermal-differential cells, 255
Thermodynamics, 111, 122
Thin film, 138
Thiols, 168
Thorazine, 41
Thrillseeking, 94
Top-down methods, 171
Top-down techniques, 102
Transferring genes, 154
Transistor, 118, 164, 269
Transmission-electron micrograph, 140
Transportation, 248
Triskaidekaphobia, 97
Trojan Horses, 258
Try to stop us, 160
Trypanosoma cruzi, 88
Tunneling, 119
Turbulent winds, 244

U

Ultrasensitive detectors, 274
Unbounding the Future, 68
Unfathomable numbers, 273
Uploading, 104, 208
Upside benefits, 158
Urticaria, 95
User communication, 129
Utopian ideals, 102

V

Valium, 41
Vapour deposition techniques, 138
Venesection, 20
Video cameras, 253
Vis medicatrix naturae, 12
Viscosity, 133
Vitamins, 110

W

Wagerphilia, 94
Weblogs, 222
Wilderness training, 258
William Cheselden, 21
Wing-warping, 243
Workaholics, 94

X

Xenophobia, 97

Y

YBCO, 140

Z

Zeptotechnology, 116
Zoophobia, 97